ORGANIC PHOTOCHEMICAL
SYNTHESES

ORGANIC PHOTOCHEMICAL SYNTHESES

VOLUME 1

1971

EDITOR

R. SRINIVASAN

IBM WATSON RESEARCH CENTER
YORKTOWN HEIGHTS, NEW YORK

ASSOCIATE EDITOR

T. D. ROBERTS

UNIVERSITY OF ARKANSAS
FAYETTEVILLE

with an
Introductory Section on
Experimental Methods in
Organic Photochemical Syntheses
by
R. Srinivasan

WILEY-INTERSCIENCE, a Division of John Wiley & Sons, Inc.
New York · London · Sydney · Toronto

Copyright © 1971, by John Wiley & Sons, Inc.

All rights reserved. Published simultaneously in Canada.

No part of this book may be reproduced by any means, nor transmitted, nor translated into a machine language without the written permission of the publisher.

Library of Congress Catalogue Card Number: 78-161147

ISBN 0-471-81920-4

Printed in the United States of America.

10 9 8 7 6 5 4 3 2 1

CONTRIBUTORS

V. Y. Abraitys
Edward Alexander
D. R. Arnold
Thomas J. Barton
Peter Beak
F. T. Bond
E. N. Cain
O. L. Chapman
J. Cornelisse
J. K. Crandall
Stanley J. Cristol
R. L. Cryberg
H. C. Cuts
Wendell L. Dilling
Richard A. Dybas
G. L. Eian
A. C. Fabian
Donald G. Farnum
C. S. Foote
R. S. Foote
C. Frater
P. D. Gardner
P. G. Gassman
A. H. Glick
Zeev Goldschmidt
A. A. Griswold
Gerald W. Gruber
D. L. Heywood
H. L. Jones
Andrew S. Kende
Grant R. Krow
W. W. Kwie

R. S. H. Liu
C. L. McIntosh
Clelia W. Mallory
Frank B. Mallory
S. Masamune
C. F. Mayer
P. de Mayo
John L. Meisel
Abdol J. Mostashari
Albert Padwa
Leo A. Paquette
D. S. Patton
J. Peyman
Martin Pomerantz
Charles E. Reineke
R. R. Sauers
L. Scerbo
W. Schinski
D. I. Schuster
B. R. Sckolnick
B. A. Shoulders
B. Sickles
L. Skattebøl
Robert L. Snell
F. I. Sonntag
Gavin G. Spence
R. Srinivasan
Edward C. Taylor
D. J. Trecker
G. Uhde
H. Westberg
E. A. Williams

R. W. Yip

PREFACE

One of the important functions that organic photochemistry can have now—and certainly in the future—is to provide the synthetic organic chemist with useful pathways to compounds that he needs as intermediates. The reduction of preparative photochemical techniques (which at present are largely known only to "practitioners" of organic photochemistry) to common practice will become possible if many of the fascinating reactions that exist in the literature are rewritten so that a nonphotochemist with a reasonable degree of manipulative skill can carry out these reactions and expect to get the reported yield. This volume, which is intended to be one of a series, was undertaken with this objective in mind.

The main part of the volume consists of detailed descriptions of thirty-nine photochemical syntheses. Some of them were selected because they are illustrative of a route to a class of useful compounds; others describe a very characteristic photochemical transformation; yet others give recipes to novel structures which are available only by photochemical routes and whose chemistry is of general interest. All of the reactions have been run many times and nearly all have been checked independently.

Since this is the first volume in this series, a section on "Experimental Methods in Organic Photochemical Syntheses" is also included. This part of the book is meant to be a general guide to photochemical methods as used in synthesis.

It is hoped that, if there is sufficient acceptance for this volume, this series will be continued. Scientists are invited to submit manuscripts to the editors for future volumes describing photochemical syntheses that they have discovered or have adapted from the literature.

R. SRINIVASAN
T. D. ROBERTS

Yorktown Heights, New York
Fayetteville, Arkansas
March 1971

CONTENTS

I. **Experimental Methods in Organic Photochemical Syntheses** 1

II. **Organic Photochemical Syntheses** 19

3-Acetoxytetracyclo[3.2.0.0^{2,7}.0^{4,6}]heptane 21
Benzonorcaradiene 23
Benzo[f]quinoline-6-carbonitrile 25
2,3-Benzotricyclo[6.1.0.0^{4,9}]nona-2,6-dien-5-one 27
Bicyclo[3.2.0]hept-6-en-3-one 28
Bicyclo[2.1.1]hexane 31
Bicyclo[2.1.1]hexane-2-one 33
N-n-Butylpyrrolidine 36
Cyclobutene 39
1β,5-Cyclo-5β,10α-cholestan-2-one (Lumicholestenone) 42
cis,trans-1,3-Cyclooctadiene 44
Dimer of 2-Aminopyridine 46
5,5-Dimethyl-1-vinylbicyclo[2.1.1]hexane 48
4,5-Diphenyl-1-methylimidazole 50
4,4-Diphenyl-3-oxatricyclo[4.2.1.0^{2,5}]nonane 51
Ethyl 1-Hydroxycyclohexanecarboxylate 53
3-Fluorophenanthrene 55
3,4-Hexadienoic Acid 58
3-Hydroperoxy-2,3-dimethyl-1-butene 60
Isophorone Dimers 62
6-epi-Lumisantonin 65
3-Methoxy-4-azatricyclo[3.3.2.0^{2,8}]deca-3,6,9-triene 67
2-Methoxy-5-hydroperoxy-2,5-dimethyldihydrofuran 70
trans-9-Methyl-1,2,3,4,4a,9a-hexahydrocarbazole 72
N-Methyl-4-nitro-o-anisidine 74
3-Methyl-2-oxatetracyclo[4.2.1.0^{3,9}.0^{4,8}]nonane 76
2-Methylenebicyclo[2.1.1]hexane 77
trans-β-Nitrostyrene Dimer 79
Norbornene Dimer 81
Pentacyclo[4.4.0.0^{2,5}.0^{3,8}.0^{4,7}]decane-9,10-dicarboxylic
  Acid Anhydride 83
anti-Pentacyclo[5.3.0.0^{2,5}.0^{3,9}.0^{4,8}]decan-6-ol 85
2-phenylbicyclo[1.1.1]pentanol-2 87
5-Phenylphenanthridine-6[5H]-one 89
Photochemical Wolff Rearrangement 92
Tetracyclo[3.2.0.0^{2,7}.0^{4,6}]heptane-2,3-dicarboxylic Acid 94
Tetracyclo[3.2.0.0^{2,7}.0^{4,6}]heptane (Quadricyclene) 97

Contents

1,3,5,7-Tetramethylcyclooctatetraene 99
Tricyclo[3.3.0.0^{2,6}]octane 101
α-Truxillic Acid 103

Index 105

ORGANIC PHOTOCHEMICAL SYNTHESES

I

Experimental Methods in Organic Photochemical Syntheses

R. Srinivasan

I. General Considerations
   A. Introduction
   B. The Crucial Factor: Time of Photolysis
   C. Calculation of Approximate Time of Photolysis
II. Experimental Methods
   A. Choice of a Light Source
      1. Choice of Wavelength
         a. Absorption by Reactant
         b. Absorption by Product
      2. Mercury Lamps
         a. Resonance Lamps
         b. Medium and High Pressure Arc Lamps
      3. Sources of Visible Light
      4. Filters
   B. Apparatus
      1. External Lamps
      2. Internal or Immersion Lamps
   C. Reaction Details
      1. Choice of Solvent
      2. Concentration
      3. Reaction Volume
      4. Outgassing
      5. Safety
      6. Temperature Control
      7. Flow Systems
      8. Gas Phase Reactions
      9. Solid Phase Reactions
      10. Sensitization
III. Bibliography
IV. References

I. General Considerations

A. Introduction. A photochemical reaction is brought about by the promotion of a molecule M to an excited state M* either by the absorption of a quantum of light

$$M + h\nu \rightarrow M^* \tag{1}$$

or by the transfer of energy from a sensitizer S in its excited state

$$S^* + M \rightarrow S + M^* \tag{2}$$

followed by the decomposition of M* to a product or products

$$M^* \rightarrow \text{product(s)} \tag{3}$$

B. The Crucial Factor: Time of Photolysis. The suitability of photochemical reaction for organic synthesis is based on not only all the factors that are applicable to a thermochemical reaction (e.g., ready availability of starting materials, moderate reaction conditions, half-life of 1–10 hours, good yield, easy purification of product) but involves, in addition, a crucial factor: excitation of a sufficient number of molecules to their reactive states in a reasonable time to obtain the required yield of the product.

An elaborate discussion of photochemical theory is beyond the scope of this volume, but a few factors which pertain to syntheses would bear examination here. The efficiency of reaction (3) determines the yield of the product but, since there is no direct way to measure the concentration of M* that is produced, the usual measure of photochemical efficiency is the *quantum yield* which is given by

$$\text{quantum yield } (\Phi) = \frac{\text{number of molecules of product}}{\text{number of photons absorbed}} \tag{4}$$

If M* is produced by sensitization (reaction 2), the quantum yield is based on the number of photons absorbed by the sensitizer.

Photon energy is measured as number of photons of a given wavelength absorbed in unit time per unit volume. One mole of photons (6.024×10^{23} photons) is equal to one *einstein*. The power (w) in watts of a light source at a given wavelength is related to the photon flux by the formula

$$\text{photon flux} = \frac{w}{E} = \frac{w\lambda}{hc} \tag{5}$$

where E is the energy of the photon, λ is its wavelength, c is the velocity of light, and h is Planck's constant.

An example will demonstrate how formulas (4) and (5) can be used to calculate the time needed to photolyze a given quantity (*g* in grams) of a reactant to give the required product:

m = molecular weight of reactant
w = output of the light source in watts *at the wavelength absorbed by the reactant*
λ = wavelength of the useful radiation in nanometers

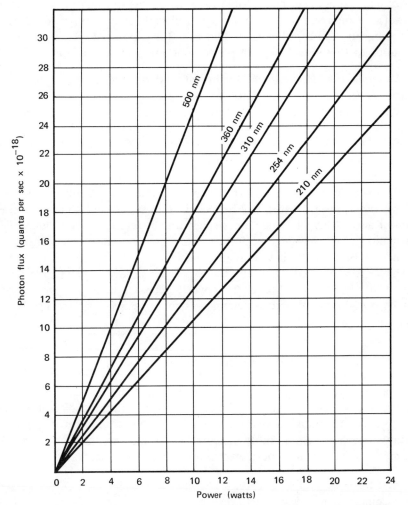

Fig. 1. Conversion chart: *power in watts to photon flux at various wavelengths.*

Φ = quantum yield for the decomposition of the reactant
t = time in seconds

$$t = \frac{g}{m} \frac{hc \times 6.024 \times 10^{23}}{w\lambda\Phi} \tag{6}$$

C. Calculation of Approximate Time of Photolysis. In order to reduce the labor involved in the use of equation (6), reference may be made to Figs. 1 and 2. If the output of the lamp is known in the wavelength region in which the reactant absorbs (the output of ultraviolet light of commonly used light sources is given in Table I), it is possible to calculate the photon flux (in photon/sec) at the average wavelength of the absorption region by using Fig. 1, From this value and Fig. 2 it is possible to read the time required to photolyze 1 mole of product for any assumed quantum yield.

Example. If a Hanovia A 673 lamp with a Pyrex filter is used to irradiate a carbonyl compound $\lambda_{max} \sim 300$ nm, the useful output of the lamp is 33 watts, and this is centered at 310 nm. From Fig. 1 the photon flux which corresponds to 3.3×10 watts is 4.8×10^{19} quanta/sec. If this value is carried to Fig. 2, it can be seen that in a reaction with a quantum yield of 0.1 it is possible to convert 1 mole of reactant in 34.9 hours.

In these calculations it is assumed that *all* the output of the lamp at the desired wavelength is available to bring about the photochemical reaction. For various reasons, which will be examined in subsequent sections, this is almost never true. The quantum yield is usually less than unity unless the formation of the product occurs by a free radical chain mechanism. A glance at Fig. 2 will show that the half-lives of photochemical reactions (which are controlled wholly by the experimental arrangement) are measured in days if not in weeks! It is worth repeating that the pumping of sufficient light energy into a reacting system is the slowest step in photochemical syntheses.

II. Experimental Methods

The practical considerations that enter into the development of a photochemical reaction to a synthetically useful method are the choice of a light source, the apparatus necessary to pump the requisite number of photons into the reactant(s), and other factors in the irradiation and workup of the sample. Each of them is considered in order in this section.

A. Choice of a Light Source. The choice of a light source is dictated by a combination of wavelength distribution, intensity, and filters.

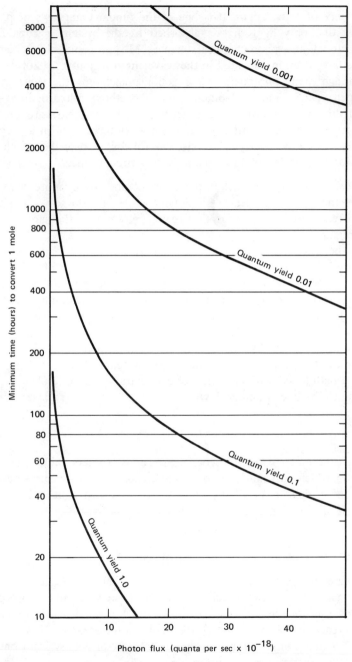

Fig. 2. Conversion chart: *Photon flux to minimum conversion times.*

1. *Choice of Wavelength.* It is one of the fundamental laws of photochemistry that only light that is absorbed by the system is capable of bringing about any photochemical change. Most organic photochemical reactions have been investigated in the wavelength region from 200 to 380 nm, although many photooxidation reactions (including the syntheses of 3-hydroperoxy-2,3-dimethyl-1-butene and 2-methoxy-5-hydroperoxy-2,5-dimethyl dihydrofuran which are described in this volume) are brought about by visible light. In order to carry out a photoreaction on a preparative scale, it is not necessary to exercise careful control over the wavelength distribution of light except under the following circumstances.

a. *Absorption by reactant.* If the reactant absorbs in more than one region of the spectrum (nearly all organic molecules do so), the photochemical reactions in one absorption region may not be the same as in a second region.

Since the different absorption regions may be quite broad, and since the absorption intensities may be widely different, the effect of using polychromatic radiation instead of a monochromatic source is not predictable. It is usual to restrict the absorption to the longest wavelength absorption band by the use of an appropriate combination of the light source and filter. In a bimolecular photochemical reaction, the same problem may arise from the absorption of light by the chromophores in each reactant. If I_0 and I are the incident and transmitted intensity of a beam of light which passes through a path length l of a solution of concentration C (in moles/l) then, according to the Beer-Lambert law:

$$\frac{I}{I_0} = 10^{-\epsilon c l} \qquad (7)$$

where ϵ is the molar extinction coefficient (units of liters/mole-cm). In a solution in which two materials of molar extinction coefficients ϵ_1 and ϵ_2 are present in concentrations C_1 and C_2, the absorbed light at any given wavelength is partitioned among the two species in the ratio:

$$\frac{\text{photons absorbed by species 1}}{\text{photons absorbed by species 2}} = \frac{\epsilon_1 C_1}{\epsilon_2 C_2} \qquad (8)$$

A suitable choice of the values of C_1 and C_2 can hence make it possible to irradiate a mixture of two species selectively even when their absorption spectra overlap for the most part, if their extinction coefficients are similar or, if the reactant that must absorb the light has the greater ϵ. It should be pointed out that complications due to the transfer of energy from one reactant to the other may also interfere.

b. *Absorption by product.* If the product absorbs some of the light from a polychromatic source, it may undergo secondary decomposition and hence lower the chemical yield of the reaction. This effect can be eliminated by the use of radiation of more restricted wavelength distribution. In cases in which the product absorbs exactly in the same region as the reactant, it will act as an inner filter and will progressively slow the reaction. In the special instance in which the product absorbs in the same region as the reactant and undergoes photodecomposition to give the reactant, the reaction will reach a photostationary state, which will be time-independent.

In solution, vibrational relaxation processes compete so efficiently with photochemical processes that irradiation at any energy level within the vibrational manifold of a single electronic state leads to nearly identical results. For preparative work the wavelength spread (width at half intensity) in the light source need not be any narrower than 50 nm. This is the approximate width of a single absorption band in solution. Fortunately for the photochemist, it is possible to obtain light output of this degree of selectivity from a combination of readily available light sources and filters.

2. *Mercury Lamps.* Over the entire near ultraviolet region from 200 to 400 nm, mercury arc lamps have been used almost exclusively as light sources. These fall into two categories.

a. Resonance lamps. These lamps emit as much as 90% of their ultraviolet output in a narrow region around 253.7 nm. There are both hot-cathode and cold-cathode types in these lamps. The former are exactly the same as the familiar fluorescent lamps used in room lighting, but the glass envelope is replaced by one of quartz and without the fluorescent coating. The output of a typical lamp is shown in Figure 3a. They have relatively long lives and usually do not need special cooling. The cold-cathode lamps operate at voltages greater than 110 volts. A starting voltage of 700 volts and an operating voltage of 300–400 volts is common. By coating the inside of a hot-cathode resonance lamp with a fluorescent coating, the primary emission at 253.7 nm can be converted into a secondary (fluorescent) emission that falls in a wavelength region of use to a photochemist. The emission of two such lamps are shown in Fig. 3, b and c. A combination of the three lamps, along with the appropriate power supply, can provide a convenient light source from 250 to 380 nm for preparative photochemistry. In some mercury resonance lamps the envelope is made of a special quartz which is quite transparent at 180 nm. These lamps are rich sources of the mercury resonance radiation at 185 nm, the output at this wavelength being about one-third of the output at 254 nm lamps. This region of the ultraviolet spectrum has hardly been exploited for preparative photochemical purposes.

b. Medium and high-pressure mercury arc lamps. These lamps operate

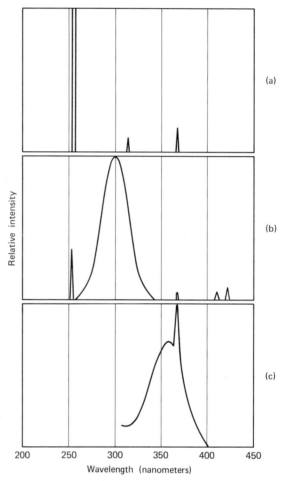

Fig. 3. Output of low-pressure mercury lamps (courtesy of Southern New England Ultraviolet Company). (a) Hot-cathode mercury resonance lamp (RUL-2537). (b) Hot-cathode mercury resonance lamp coated with "Sunlight Phosphor" (RUL-3000). (c) Hot-cathode mercury resonance lamp coated with "Blacklight Phosphor" (RUL-3500).

at internal pressures from 1 to 100 atm. Two features which make them very useful in preparative organic photochemistry are their broad spectral emission and their compactness.

The output of a medium-pressure arc shows many emission lines over the whole spectrum from 200 to 600 nm. In a high-pressure arc the broadening of the lines gives rise to nearly continuous emission in the visible region (accounting for the "whiteness" of the light), but there is some loss in intensity in the ultraviolet through reabsorption.

Table I. Output of ultraviolet light from commercial mercury lamps for photochemical syntheses on laboratory scale

Lamp type	Manufacturer	Catalog number	Operating voltage	Input (watts)	Approximate life (hours)	Ultraviolet light output (watts) 380–330 nm	330–280 nm	280–220 nm	Remarks
1. Low-pressure (Hg resonance)									
a. Cold-cathode	Westinghouse[a]	WL-782-20	325	14	—	—	—	2.0[c]	Output in other spectral regions ∼0
b. Hot-cathode	General Electric Co.[b]	G25T8	42	25	7500	—	—	5.0[c]	Output in other spectral regions ∼0
	Southern New England Ultraviolet Co.	RUL-2537	—	100	3000	—	—	15[d]	See Fig. 3 for output in other regions
		RUL-3000	—	100	3000	—	11[d]	—	
		RUL-3500	—	100	3000	13[d]			
3. Medium-pressure	Engelhard[e] Hanovia, Inc.	679A-36	135	450	1000[h]	28.0	28.7	27.0	
		673A-36	145	550	1000[h]	32.9	32.8	29.2	
4. High-pressure	General Electric Co.[f]	AH-6[g]	840	1000	75[h]	80	56	33	Water cooling essential

[a] Westinghouse data sheet ACS-142.
[b] General Electric Technical Publication LS-179.
[c] Very narrow band centered at 253.7 nm.
[d] Per module.
[e] Bulletin EH-223 Hanovia Lamp Division.
[f] General Electric Company Bulletin GET-1248J.
[g] Quartz water jacket.
[h] These values are greatly dependent on the operating conditions and the number of starts.

Both the medium- and high-pressure arcs operate at temperatures considerably greater than the ambient room temperature. The lamps of wattage greater than 100 watts require external cooling by air or circulating water especially when they are operated in a confined volume. Overcooling can prevent these lamps from reaching their full intensity, and undercooling can shorten their lives considerably and even cause them to explode.

It should be noted that, although medium- and high-pressure mercury arcs have a considerable part of their ultraviolet output in the 254 nm region, the center of the mercury resonance line is "reversed," *i.e.*, it is removed from the emission by self-absorption in the lamp itself. As a result these lamps cannot be used in mercury photosensitization reactions.

In Table I the operating characteristics of typical examples of each of these lamps are listed. All of them are suitable for photochemical synthesis on a laboratory scale. A summary of data on more powerful ultraviolet light sources for syntheses on a scale two- to tenfold larger may be found elsewhere.[1]

3. *Sources of Visible Light.* For preparative photochemical work with visible light a number of light sources are available. Included are the high-pressure mercury arcs mentioned before, tungsten lamps, tungsten-halogen lamps, and carbon arcs. The characteristics of some convenient sources of visible light are given in Table II.

Table II. **Lamps for photochemical syntheses with visible light**[a]

Manufacturer	Catalog number	Operating voltage	Nominal watts	Intensity[b] (quanta/min)
Sylvania[c]	DWY Tungsten-Halogen	117	650	29.5×10^{20} [d]
	DWY Tungsten-Halogen	70		9.6×10^{20}
Sylvania[c]	500 Q/Cl	117	500	16.8×10^{20}
	500 Q/Cl	70	—	3.0×10^{20}

[a] Intensity data kindly supplied by Professor C. S. Foote, University of California, Los Angeles.
[b] Actual intensities absorbed by 10^{-4} M Rose Bengal.
[c] Equivalent lamps are available from the General Electric Company.
[d] Believed to be a minimum value.

Sunlight is a rich source of radiation for photochemical work. Its use is naturally restricted to tropical and subtropical regions, but its practicality has been demonstrated by work carried out in Italy, Egypt, and California. The light that reaches the earth's surface is a mixture of direct sunlight and light scattered from the sky. The total radiation has a sharp cutoff in wavelength at about 310 nm. The intensity peaks at about 480 nm and decreases very gradually toward the long wavelength end of the visible spectrum. In using sunlight for preparative photochemical reactions, it should be emphasized that (a) it is best suited for reactions which occur in visible light; (b) its intensity is approximately 1/25th of that of a xenon arc (1000 watts) at 5 in.; and (c) it can be used to maximum advantage only if the sample being irradiated is moved to follow the changing direction of the sun's rays during the course of a day.

4. Filters. To restrict the wavelength of the light emitted by medium- and high-pressure mercury lamps, it is customary to use a filter between the lamp and the reactant. These filters usually cut off the light at the short wavelength end of the spectrum. Typical filters and their transmission characteristics are given in Table III.

Table III. Transmission characteristics of filters for photochemical syntheses

Material	Nominal thickness (mm)	Transmission (%)			
		$\lambda = 254$ nm	300 nm	360 nm	450 nm
Quartz[a]	2	90	90	90	90
Vycor 791	2	90	90	90	90
Corex D[b]	2	0	60	90	90
Pyrex[c]	2	0	35	90	90
Window glass	2	0	5	90	90

[a] Clear fused from General Electric Company.
[b] Corning 9–53.
[c] Corning 0–53.

Ultraviolet lamps sometimes incorporate a filter material as an envelope for the lamp. It is important to obtain from the manufacturer the information concerning the wavelength distribution of each lamp.

The output of mercury resonance lamps (with or without fluorescent coating) is sufficiently narrow for preparative work. Hence filters are not needed with these lamps.

When visible light from artificial sources or from the sun is used to carry out a reaction, it is sometimes necessary to filter out the light in the near ultraviolet region. Soft glass is a good material for this purpose (Table III).

B. Apparatus. Organic syntheses in the laboratory rarely exceed a scale of 1 mole. Therefore the following discussion of equipment for photochemical processes is restricted to operations on this scale or smaller.

There are two principal arrangements for carrying out photochemical reactions. In one, the light source (or sources) is placed outside the vessel which contains the sample. In the second, the situation is reversed and the sample surrounds the light source. The equipment used in the latter case is commonly referred to as an "immersion unit." Since both geometries have advantages and disadvantages, it is better to consider the two classes of equipment separately.

1. External Lamps. It is obvious that any source of light can be placed outside a reaction vessel to bring about a photochemical reaction. The most important consideration is that the material of the vessel should be transparent to the radiation that is meant to bring about the change. Cooling the sample may be a significant point when medium- and high-pressure mercury arc light sources are used. For photoreactions in the visible region of the spectrum, where photons are plentiful and cheap, this arrangement is generally used.

A typical commercial reactor in which the sample is placed at the center of a circular bank of lamps is shown in Fig. 4. The lamps in such reactors are invariably low-pressure mercury sources which are available with or without fluorescent coatings. Such reactors can be used for reactions at 254, 300, and 360 nm by the use of appropriate sets of lamps.

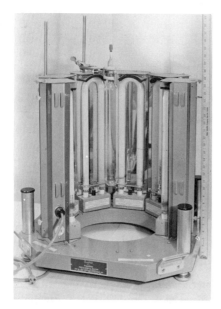
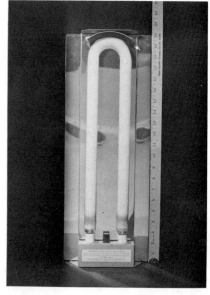

Fig. 4. Rayonet Type RS preparative photochemical reactor, model RPR-208 (manufactured by Southern New England Ultraviolet Company, Middletown, Conn.) The cover and two of the modules have been removed to show the reaction chamber. The inset shows a single module. The scale is calibrated in inches.

The efficient utilization of the photon flux from the lamps requires that the sample be contained in a volume whose length is the same as that of the radiant part of the lamps. It follows that, to accommodate samples of differing volumes, the lengths of the reaction vessels should be kept constant and the diameter should be changed. The photon flux is a maximum along the vertical axis of these reactors and decreases gradually toward the periphery.

The large irradiated volume in such an arrangement permits the use of vessels in a variety of shapes and sizes. These factors are of particular importance in the irradiation of dilute solutions of samples or of vapor samples on a preparative scale.

Since low-pressure mercury lamps operate best at about 60°, there is no difficulty in cooling the sample in this arrangement. Unless the top of the reactor is covered, the sample can be kept at 40° or less by means of a fan that can be located at the bottom of the reactor. Samples which have to be cooled to lower temperatures can be irradiated in a vessel fitted with a cold finger that is cooled by water or Dry Ice.

A significant advantage in this arrangement is the relatively large surface area of the reaction vessel that is exposed to light. Since almost all photochemical reactions lead to some condensed material as a side product which deposits on the wall of the vessel, the useful light that is availabe for the reaction decreases with time. It is hence advantageous to have a large surface area through which light is admitted to the sample.

The chief disadvantage in this arrangement is that the light from the lamps is not used efficiently. Even when reflectors with a high degree of reflectivity are placed around the lamp (or lamps), this geometry leads to the loss of a large fraction of the photons emitted by the lamps. In laboratory practice, when sample sizes smaller than 0.5 mole are involved, this does not constitute a problem since the power of the lamps is adequate to give a useful *absolute* photon flux within the volume of the sample. For larger sample sizes, *unless there is some other overriding advantage,* this arrangement should be avoided.

2. *Internal or Immersion Lamps.* The principal requirement for immersion units used in photochemical reactions is that the radiant part of the light source be compact so that it can be immersed completely in the sample. A low-pressure mercury arc (which has a low luminosity per unit surface compared to a medium- or high-pressure arc) is usually coiled into a tight spiral for this purpose. Though a hot-cathode, low-pressure mercury arc can be used as an immersion source, there appears to be no commercial unit that utilizes such a lamp.

The most generally used immersion units are those that are designed to accept a medium- or high-pressure mercury arc lamp. A typical commercial unit is shown in Fig. 5. Since the light source is located in a confined area, the sample is protected from the heat emitted by the lamp by a layer of coolant, usually water.

The principal advantage in this design is the efficient use of the entire photon output of the source. As medium- and high-pressure mercury light sources are available in power ratings up to many kilowatts, these immer-

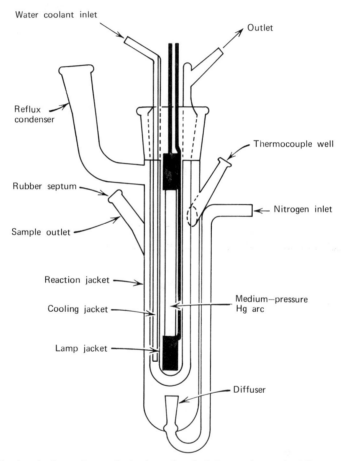

Fig. 5. Sectional view of a typical photochemical immersion unit (diagram kindly supplied by Dr. D. J. Trecker, Union Carbide Corporation). In addition to ports for cooling, this apparatus contains ports for purging with nitrogen, fitting a reflux condenser, and sampling the system. A thermocouple may be used to monitor the temperature of the reactant.

sion lamps are uniquely suited for organic photochemical syntheses on a scale of 0.5 mole or more.

As noted in Section A, medium- and high-pressure mercury arcs have a broad spectral output which extends over the range 200–600 nm. Hence immersion units can be used for irradiating samples over this whole range. However, the use of appropriate filters (usually in the form of hollow tubes which are placed around the lamp) may be necessary, as discussed earlier.

The need for continuous and intensive cooling in immersion units which use medium- (or high-) pressure mercury arcs, can be a source of operational problems for several reasons. If the coolant is water, it should be of sufficient purity to prevent the absorption of some of the useful radiation in the ultraviolet. A sudden failure in the flow of the cooling water can result in the loss of the sample and breakage of the immersion well and of the lamp in some cases. A fire can follow if any volatile organic material comes in contact with the hot lamp. In order to avoid such accidents, it is desirable to include a switch in the cooling-water line which will turn off the lamp whenever the rate of flow of water drops below a preset value.

In a unit such as that illustrated in Fig. 5 the sample is held in the annular space between two concentric cylinders. The irradiation of small (<100 ml) samples can be awkward with this arrangement.

In the second section of this volume, in which directions for the synthesis of many compounds are given, it can be seen that in several instances the authors and the checkers of a given preparation have used different equipment and even different geometries for conducting the irradiation. Such procedures demonstrate that there is no unique apparatus for carrying out a photochemical synthesis except in some unusual cases. Once the conditions for an irradiation have been worked out, the equipment that is available can be adapted to the reaction by suitable modifications in the scale of the operation and the concentration of the solution. A huge investment in lamps and reaction vessels can be avoided in this manner.

C. Reaction Details. In this section the operational details that follow the selection of a light source and a reaction vessel are considered.

1. Choice of Solvent. The medium in which a reaction is carried out can be solid, liquid, or vapor, although the vast majority of organic photochemical syntheses are conducted in the liquid phase. Since the use of the reactant in a pure liquid state is of limited interest, an important factor is the choice of a suitable solvent. Unless the reaction is one in which the solvent takes part chemically, the solvent should be as unreactive as possible. In addition to inertness in the usual thermal sense, solvents for irradiations should be transparent in the spectral region of interest. Equation (8) can be used to calculate that, if the solute is to be photolyzed until its concentration drops to 1% that of the solvent, then the extinction coefficient of the solute should be at least 5000 times larger than the extinction coefficient of the solvent at that wavelength. The choice of a solvent can be readily made on this basis. Particular care should be taken to use a spectroscopically pure grade of solvent, when the irradiated solution is even initially very dilute (1 or 2%) with respect to the solute. The other extreme of using only spectroscopically pure solvents all the time is also unnecessary in preparative photochemistry. Any impurity in the solvent, if its concentration is about 1%, *may not* interfere with the reaction so long as the radiation is being absorbed exclusively by the reactant.

2. Concentration. Photoisomerization reactions are concentration-independent and should be run in as dilute a solution as possible to minimize bimolecular side reactions (*e.g.,* dimerization). On the other hand, bimolecular reactions are promoted by an increase in the concentration of the addend and should be run at as high a concentration of the latter as possible. More complex situations are possible as in the photocycloaddition of a carbonyl compound to an olefin. Oxetane formation will be promoted by an increase in the olefin concentration, but sensitized dimerization of the olefin, if it also occurs, will increase even faster. In such complex systems the optimum concentration (which can depend on the photon flux as well) can be determined only by trial.

3. Reaction Volume. If the reactant has an extinction coefficient of 1000 liters mole^{-1} cm^{-1} and its concentration is 0.1 M, then 95% of the useful radiation will be absorbed in a layer of solution that is only 0.13 mm thick. This intense concentration of photons in a thin layer in contact with the wall through which the light enters the reaction vessel will result in concentration effects which are especially undesirable in bimolecular photochemical reactions as discussed in Section 2. If the reaction cannot be run at a lower concentration, some means of stirring the solution is essential. Bubbling a stream of nitrogen through the reaction mixture or stirring with a magnetic stirrer may be adequate. Reaction vessels in which the solution is held in a narrow space (*e.g.,* in an immersion unit) would need more vigorous stirring than those which are in the form of cylinders or spheres.

4. Outgassing. For preparative work, outgassing or purging of the solution is generally unnecessary. The concentration of oxygen in a typical solvent (*e.g.,* benzene) is about 10^{-3} M. As long as the reaction vessel is kept closed, this oxygen will be used up in the initial stages of the photoreaction. The beneficial effect of purging the solution with nitrogen during irradiation lies mainly in the mixing effect that is realized.

5. *Safety.* The hazards that are peculiar to photochemical equipment are due to the following factors.

a. Intense ultraviolet radiation. Tinted glasses for the eyes and light gloves for the hands usually give adequate protection. A light baffle of cardboard or aluminum film is sufficient to cut off any exposure to reflected ultraviolet light. Good ventilation is necessary to sweep away the ozone that may build up around the lamps.

b. High voltage in the light sources. Medium- and high-pressure mercury arcs operate on voltages which can range from several hundred to several thousand volts. The combination of fragile lamps, water cooling, and high voltages makes safety precautions such as grounding of all the wiring in the equipment absolutely essential.

c. Fire hazard. Since photochemical reactions are continuously run for days and frequently in the absence of the operator, it is important to anticipate emergencies which can result from a failure in the water supply, cooling air, or the cracking of glassware under stress. Isolation of the equipment in a well-ventilated hood is a minimum requirement.

6. *Temperature Control.* Photoreactions in the condensed phase show very little sensitivity to changes in temperature of ±30°. Control of the reaction temperature within these limits is of interest mainly to prevent the loss of the reactant(s) or solvent by evaporation.

7. *Flow Systems.* Photochemical reactions in which the desired product(s) is sensitive to further photolysis should show improved efficiency when the reaction is conducted in a flow system. Descriptions of elaborate arrangements to achieve this in small-scale work can be found in the literature.[2] There appears to be no simple apparatus that can be readily attached to commercial irradiation units to achieve the same end.

8. *Gas Phase Reactions.* Though many striking photoreactions have been described in gas phase studies, the main difficulty in adapting most of them to preparative work is the small amount of material that can be handled in a gas system, especially at a pressure less than 1 atm. Once again, a flow system will provide a solution, but none is available commercially. One solution to this problem is to run the reaction at 1 atm as described in the synthesis of bicyclo [2.1.1] hexane in this volume. A system which operates at the boiling point of the reactant can feed the vapor continuously to the reaction chamber, and the mixture of product and unreacted materials can be returned to the pot. Unless a fractionation unit is included in the train, this arrangement leads to considerable secondary

photolysis of the product. It should also be pointed out that many vapor phase reactions may be quenched out at 1 atm pressure.

9. Solid Phase Reactions. Certain photodimerizations seem to proceed very efficiently in the solid phase. In the irradiation of solids the penetration of light beyond a few surface layers presents a problem. The use of finely powdered material, which is constantly turned to expose a fresh surface to the light beam, is a commonly adopted technique. The powdered material can also be handled as a slurry or a suspension. (See synthesis of α-truxillic acid)

10. Sensitization. The theory of the photosensitization of a chemical transformation (reaction 2) has been dealt with in detail in many books. Assuming that a given reaction can be photosensitized, the factors that need to be considered are the following.

a. The sensitizer should be relatively unreactive in the environment. It should not react with itself, the solvent, or the product.

b. It should absorb the light to the virtual exclusion of light absorption by the acceptor or product.

c. It should have an excited state energy within 10 kcal of the acceptor, preferably higher in energy than the acceptor.

d. The product should not readily accept energy from the sensitizer.

Triplet energies of commonly used sensitizers have been given in the literature.[3]

Bibliography

L. R. Koller, *Ultraviolet Radiation,* 2nd ed., John Wiley & Sons, New York, 1965.

W. A. Noyes, Jr., and P. A. Leighton, *Photochemistry of Gases,* Rheinhold Publishing Corp., New York, 1941.

J. G. Calvert and J. N. Pitts, Jr., *Photochemistry,* John Wiley & Sons, New York, 1966.

References

1. "*Industrial Photochemistry,*" *Ind. Eng. Chem.* (8) **54,** 28 (1962).
2. G. O. Schenck, in A. Schonberg, ed., *Praparative Organische Photochemie,* Springer-Verlag, Berlin, 1958.
3. W. G. Herkstroeter, A. A. Lamola, and G. S. Hammond, *J. Amer. Chem. Soc.,* **86,** 4537 (1964).

II

Organic Photochemical Syntheses

3-ACETOXYTETRACYCLO[3.2.0.0^{2,7}.0^{4,6}]HEPTANE

Submitted by P. G. GASSMAN and D. S. PATTON*
Checked by M. POMERANTZ and G. W. GRUBER†

1. Procedure

The apparatus consists of a 700-ml irradiation vessel equipped with a water-cooled internal quartz immersion well (Note 1) containing a 450-watt Hanovia medium-pressure lamp and fitted with provisions for maintaining a slow stream of nitrogen through the solution, and for magnetic stirring. In this vessel are placed 3.00 g (0.02 mole) (Note 2) of 7-acetoxybicyclo[2.2.1]heptadiene (Note 3) and 600 ml of pentane (Note 4). The solution is irradiated under a nitrogen atmosphere, with stirring, for 2 hours. The irradiation is stopped and the thin film of polymer which deposits on the outer wall of the immersion well is removed (Note 1). The apparatus is reassembled and the irradiation continued for 2 hours (Note 5). The irradiation is stopped and the irradiated solution is transferred to a 3-liter Erlenmeyer flask. Saturated aqueous silver nitrate solution (10 ml) is added to the pentane solution and the mixture is stirred vigorously for ca. 0.5 hour to remove any unreacted starting material. The pentane solution is then decanted from the aqueous solution and precipitated silver complex. The nondecanted material is washed twice with 100-ml portions of pentane and the pentane solutions are combined, dried over anhydrous magnesium sulfate, filtered, and the solvent is removed by distillation at atmospheric pressure. The residue is distilled under reduced pressure to yield 2.2–2.3 g (74–77%) (Note 6), b.p. 68–70° (4 mm). On cooling, this material solidifies to give white crystalline 3-acetoxytetracyclo[3.2.0.0^{2,7}.0^{4,6}]heptane, m.p. 33–35° (Note 7).

* The Ohio State University, Columbus, Ohio 43210.
† Case Western Reserve University, Cleveland, Ohio 44106.

2. Notes

1. Before use, the quartz immersion well should be cleaned by brief rinsing with 50% aqueous hydrofluoric acid (*Caution!*), followed by thorough rinsing with water and acetone. After drying, the immersion well is ready for use.

2. When concentrations in excess of 0.1 M are used, yields and product purity diminish slightly.

3. Material obtained from Frinton Laboratories was used without purification.

4. Commercial grade pentane obtained from Matheson, Coleman and Bell was used without purification.

5. The progress of the irradiation can be followed by vpc analysis using a $\frac{1}{8}$ in $\times$ 10 ft 5% SE 30 on 80/100 Diatoport S column at 100°.

6. In a series of 14 runs on this scale using starting material prepared according to the literature procedure,[1] the yields varied from 67 to 84% with an average yield of 77%. On a larger scale, yields of 90% could be realized.[2]

7. During the distillation, the solidification of the material in the condenser can plug the system. Alternatively, the material can be purified by sublimation.

3. Methods of Preparation

The procedure described is a modification of that outlined in the literature.[2]

4. Merits of the Preparation

This procedure provides a general method for the photochemical conversion of norbornadienes into derivatives of tetracyclo[3.2.0.0^{2,7}.0^{4,6}]heptane.

References

1. P. R. Story, *J. Org. Chem.*, **26**, 287 (1961)
2. P. G. Gassman and D. S. Patton, *J. Am. Chem. Soc.*, **90**, 7276 (1968), and references therein.

BENZONORCARADIENE

Submitted by MARTIN POMERANTZ and GERALD W. GRUBER*

1. Procedure

Twenty grams (0.14 mole) of 1,2-benzotropilidene (Note 1) and 4650 ml of anhydrous ether are placed in a 5-liter three-necked flask fitted with a water-cooled Pyrex immersion well, a rubber septum, a magnetic stirring bar, and a water-cooled reflux condenser capped with a calcium chloride drying tube. The solution is stirred magnetically while irradiating with a 400-watt G.E. medium-pressure mercury arc (cat. No. H400A33-1, Note 2) and is monitored by glpc (Note 3). After 380 minutes of irradiation, analysis by glpc (Note 3) shows that the solution contains 92% of benzonorcaradiene, 5% of 1,2-benzotropilidene, a total of 3% of naphthalene, and several minor products (Note 4).

The solution is concentrated to *ca.* 400 ml with a flash evaporator, reduced to *ca.* 30 ml by distillation through a 6-in. Vigreux column, and finally fractionated with a 2-ft Teflon spinning-band column. The spinning-band distillation is accomplished with a pot temperature of 140–150°, a column temperature of 120°, and a reflux ratio of *ca.* 125:1. The distillation is monitored by glpc as described above (Note 3). After a 4.4-g forerun, b.p. 110–124° (20 mm) (Note 5), 13.6 g (68% of the theoretical yield) of pure (Note 6) benzonorcaradiene is collected at 124–125° (20 mm).

2. Notes

1. Of the several methods which have been reported[1a–i] for the preparation of 1,2-benzotropilidene the most convenient is the pyrolysis[1e–f] of benzonorbornadiene. Benzonorbornadiene is readily prepared[2a–c] by the reaction of benzyne with cyclopentadiene. The pyrolysis reaction is carried out essentially as described previously.[1f] An 18-in. Pyrex tube is packed with glass beads and placed vertically along the axis of an open-ended oven.

* Case Western Reserve University, Cleveland, Ohio 44106.

Atop the tube is placed a pressure-equalizing dropping funnel fitted with a nitrogen inlet system. The lower end is connected to a series of three Dry Ice-cooled traps. When the oven temperature reaches 530–40° benzonorbornadiene is placed in the dropping funnel and a slight positive nitrogen pressure applied. The benzonorbornadiene is then added dropwise (*ca.* 20 ml/hr), the residence time being 3–6 seconds. Glpc analysis (8 ft × ¼ in. column of 20% Carbowax 20M on 60/80 mesh Chromosorb P operated at 170° with a carrier gas flow rate of 45 ml/min) indicated 86% of 1,2-benzotropilidene, 14% of α- and β-methylnapthalenes, and only a trace of benzonorbornadiene. Pure (99%) 1,2-benzotropilidene is obtained by distillation through a 2-ft Teflon spinning-band column. [Pot 140–150°, column 120°, reflux ratio *ca.* 100:1; b.p. 123–124° (20 mm).

2. The outer Pyrex lamp housing and the base were removed from the lamp, leaving only the quartz envelope and power supply connections, which were then placed inside the immersion well.

3. A 4 meter × 0.25 in. column packed with DC 710 silicone oil on 45/60 mesh Chromosorb P operated at 180° with a carrier gas flow of 92 ml/min afforded the following relative retention times: naphthalene, 1.15; 1,2-benzotropilidene, 1.65; benzonocaradiene, 1.90. No correction was made for differences in thermal conductivities of the products.

4. Further irradiation leads to significant amounts of naphthalene at the expense of benzonorcaradiene.

5. As calculated from glpc data (uncorrected for thermal conductivity differences), this fraction contained 0.1 g of benzobicyclo[3.2.0]hept-2,6-diene, 0.5 g of naphthalene, 0.9 g of 1,2-benzotropilidene, and 2.9 g of benzobicyclo[4.1.0]hept-2,4-diene.

6. The material was greater than 98% benzonocaradiene according to analysis on two glpc columns. The only detectable impurities (*ca.* 1% total) were methylnapthalenes.

3. Methods of Preparation

Benzonorcaradiene has been prepared by photolytic[3] and cuprous bromide[1d]-catalyzed decomposition of diazomethane in the presence of naphthalene. Most recently[4] it has been prepared by the photochemically induced reorganization of 3,4-benzotropilidene.

4. Merits of the Preparation

In addition to the hazards of diazomethane, the methods of Doering and Müller lead to a mixture of products from which small amounts of pure benzonorcaradiene can be extracted only with considerable difficulty.

3,4-Benzotropilidene, upon irradiation, produces benzonorcaradiene in high yield; however, the starting material is relatively inaccessible. This preparation affords a reasonable yield of pure material, with relative ease and low cost.

References

1. (a) G. Wittig, H. Eggers, and P. Duffner, *Ann* **619**, 10 (1958); (b) R. Huisgen and G. Juppe, *Chem. Ber.*, **94**, 2332 (1961); (c) W. Ziegenbein and W. Lang, *ibid.*, **95**, 2321 (1962); (d) E. Müller, H. Fricke, and H. Kessler, *Tetrahedron Lett.*, 1525 (1964); (e) R. H. Hill and R. M. Carlson, *J. Org. Chem.*, **30**, 2414 (1965); (f) S. J. Cristol and R. Caple, *ibid.*, **31**, 585 (1966); (g) M. Pomerantz and G. W. Gruber, *J. Amer. Chem. Soc.*, **89**, 6799 (1967); (h) V. Rautenstrauch, H. J. Scholl, and E. Vogel, *Angew. Chem. Internat. Ed. Engl.*, **7**, 288 (1968); (i) W. Metzner and K. Morgenstern, *ibid.*, **7**, 379 (1968).
2. (a) L. Friedman and F. M. Logullo, *J. Amer. Chem. Soc.*, **85**, 1549 (1963); (b) T. F. Mich, E. J. Nienhouse, T. E. Farina, and J. J. Tufariello, *J. Chem. Ed.*, **45**, 272 (1968); (c) we wish to thank Professor Friedman for the preparation. L. Friedman, F. M. Logullo, and D. M. Smith, to be submitted to *Organic Syntheses*.
3. W. von E. Doering and M. J. Goldstein, *Tetrahedron*, **5**, 53 (1959).
4. M. Pomerantz and G. W. Gruber, *J. Amer. Chem. Soc.*, **89**, 6898 (1967).

BENZO[*f*]QUINOLINE-6-CARBONITRILE

$$\xrightarrow[\text{t-BuOH, benzene, oxygen}]{h\nu}$$

Submitted by RICHARD A. DYBAS*
Checked by J. N. C. HSU†

1. Procedure

To 2-phenyl-3-(2-pyridyl) acrylonitrile (Note 1) (8.00 g, 0.039 mole) in a 5-liter flask is added 475 ml of benzene and 4.4 liters of t-butyl alcohol. Oxygen is bubbled into the reaction solution for 2 hours before irradiation

---

* University of Rochester, Rochester, New York 14627.
† IBM Research Center, Yorktown Heights, New York 10598.

and continued throughout the photolysis. The solution is irradiated for 12–13 hours (Note 2) with constant magnetic stirring, after which time the solvents are evaporated (reduced pressure) to give a red-brown residue. The residue is taken up in 2 liters of benzene and passed rapidly through a column of 150 g of Woelm alumina (activity III) to remove polymeric material. The column is washed with an additional 500 ml of benzene. The effluents are evaporated (under reduced pressure) to give 6.07 g of impure nitrile as tan needles. Recrystallization from chloroform-hexane gives 4.35 g of the nitrile as off-white needles, m.p. 162–163°. The mother liquor is evaporated (under reduced pressure) to give a red-brown oily residue (1.58 g) which is chromatographed on 50 g of Woelm alumina (activity III). Elution with mixtures of hexane-benzene gives an additional 0.23 g of the nitrile, m.p. 161–163°. Two recrystallizations of the combined material give the nitrile (3.47 g, m.p. 163–164°) as fine colorless needles. Recrystallization of the mother liquors gives an additional 0.79 g of the colorless nitrile, m.p. 163–164°. The total yield of pure benzo[f]quinoline-6-carbonitrile is 4.26 g (54%) (Note 3).

2. Notes

1. The compound was prepared by the method of Castle and Seese.[1]
2. Irradiations were carried out through a water-jacketed Hanovia 450-watt quartz immersion well fitted with a Corex 9700 filter sleeve. Benzene was added to prevent freezing of the reaction solution.
3. The same irradiation carried out in benzene solution requires 60 hours and, on the exact workup, gives the pure nitrile in 53% yield.

Reference

1. R. N. Castle and W. S. Seese, *J. Org. Chem.*, **20**, 987 (1955).

2,3-BENZOTRICYCLO[6.1.0.0⁴,⁹]NONA-2,6-DIEN-5-ONE

[Structure (I)] $\xrightarrow{h\nu, \text{Sensitizer}}$ [Structure (II)]

Submitted by A. S. KENDE and Z. GOLDSCHMIDT*
Checked by T. D. ROBERTS†

1. Procedure

A solution of 1.03 g (0.00567 mole) of the tricyclic ketone (I) (Note 1) and 5.70 g (0.0313 mole) of benzophenone (as sensitizer) in 100 ml of dry acetonitrile (Note 2) is placed in the inner well of a Pyrex reaction vessel. The solution is purged with a gentle stream of nitrogen (Note 3) for a few minutes, then irradiated at room temperature under a nitrogen atmosphere using an external medium-pressure Hanovia utility lamp (Note 4) for 3 hours.

The solvent is removed at reduced pressure and room temperature, and the residue is chromatographed over a column (10 in. long, 2 in. in diameter) of silica gel (Note 5). The excess benzophenone is removed by elution with 5:1 petroleum ether-ether (Note 6) until no further amounts are eluted, after which the desired ketone (II) can be obtained by elution with 1:1 petroleum ether-ether as solvent. Removal of solvent from the latter fractions gives a tan, semicrystalline mass which on a single recrystallization from hexane gives 0.48 g (47%) of the photoketone (II) in the form of colorless needles, m.p. 110–111°.[1] The uv, ir, and nmr spectra of this material are in accord with literature data.

2. Notes

1. Preparation of the starting trienone from cycloaddition of benzyne to tropone is described by Ciabattoni and co-workers.[2] The checker used a sample supplied by the submitters.
2. The solvent was "Spectro Grade" acetonitrile, available from East-

* University of Rochester, Rochester, New York 14627.
† University of Arkansas, Fayetteville, Arkansas 72701.

man Organic Chemicals, Rochester, New York. The acetonitrile was dried for 48 hours over molecular sieves before use.

3. Standard "Dry Nitrogen," available from Airco, was used in this experiment.

4. Model 30620, nominal 140-watt rating.

5. The silica gel used was E. Merck AG (Darmstadt), 70/325 mesh (ASTM), available (cat. No. 7734) from Brinkmann Instruments, Inc., Westbury, New York 11590. The checker found that a clean separation was not obtained on Baker 3405 silica gel.

6. The petroleum ether was of reagent grade, b.p. 30–60°.

References

1. A. S. Kende and Z. Goldschmidt, *Tetrahedron Lett.*, 783 (1970).
2. J. Ciabattoni, J. E. Crowley, and A. S. Kende, *J. Amer. Chem. Soc.*, **89**, 2778 (1967).

BICYCLO[3.2.0]HEPT-6-EN-3-ONE

Submitted by D. I. SCHUSTER and B. R. SCKOLNICK*
Checked by T. D. ROBERTS†

1. Procedure

A. 3,5-Cycloheptadienone. In a 2-liter three-necked flask equipped with a mechanical stirrer, reflux condenser, and dropping funnel with a pressure-

---

* New York University, University Heights, New York, New York 10453.
† University of Arkansas, Fayetteville, Arkansas 72701

equalizing side arm is placed 2.69 g (0.284 equivalent) of lithium aluminum hydride in 380 ml of anhydrous ether. To the stirred suspension at room temperature is slowly added 15 g (0.14 mole) of tropone (Note 1) in 300 ml of anhydrous ether. The mixture is stirred for 30 minutes at room temperature after addition is completed, then 42.1 g (0.57 equivalent) of ethyl formate in 275 ml of ether is added dropwise to the reaction flask cooled in an ice-water bath. After 30 minutes the mixture is hydrolyzed by the addition of 5.4 ml of water and 4.3 ml of 10% sodium hydroxide solution. The mixture is stirred until the salts become granular, whereupon it is filtered. The salts are washed three times with ether. The combined ethereal solution is dried over anhydrous magnesium sulfate, filtered, and the solvent removed at reduced pressure on a rotary evaporator. Spinning-band distillation of the residue gives 5 g (30% of theory) of pale yellow liquid, b.p. 52–54° (9.5 mm). Analysis by gas-liquid partition chromatography (6 ft × $\frac{1}{8}$ in., DC-200 silicone on 60/80 mesh Chromosorb P, injection port 180°, column 125°, detector 200°, flow rate 60 cc/min) indicates only traces of 3,5-cycloheptadienol impurity in the 3,5-cycloheptadienone. The infrared and ultraviolet spectra of the product agree with literature data.[1-4]

B. *Bicyclo[3.2.0]hept-6-en-3-one.* The sensitized photolysis of 3,5-cycloheptadienone[5] is carried out using a Hanovia 450-watt high-pressure mercury lamp 679A 36 in a water-jacketed quartz immersion well (Hanovia No. 19434) with a Pyrex cylindrical filter sleeve, or in a similar immersion well made of Pyrex (Note 2). The reaction vessel is a cylindrical vessel with a 60/50 female joint made to fit around the immersion well, equipped with a side arm and a 24/40 female joint for a reflux condenser, and a very short side arm near the bottom equipped with a stopcock for removing small samples for analysis. There is also a gas inlet tube entering the flask near the bottom. The content of the vessel when enclosing the immersion well is about 600 ml. The solution is stirred with a magnetic stirring bar during irradiation. The progress of the reaction is followed by glpc analysis on the DC-200 column (above), injection port 235°, column 130°, detector 300°, flow rate 28 cc/min. Bromobenzene is used as an internal standard for glpc analysis and is added to each sample after photolysis. Under the conditions above, the retention times of bromobenzene, bicyclo-[3.2.0]hept-6-en-3-one, and 3,5-cycloheptadienone are, respectively, 3 minutes 26 seconds, 3 minutes 52 seconds, and 4 minutes 30 seconds.

A solution of 0.51 g of 3,5-cycloheptadienone in 500 ml of acetone is placed in the apparatus described above. Nitrogen is bubbled through the solution for 30 minutes before photolysis, and a positive nitrogen pressure is maintained during the photolysis. The lamp is fired, and the reaction allowed to proceed for 35 minutes, after which time all of the starting material has been consumed. Analysis by glpc indicates only one major product on the column described above and also on a 4 ft × ⅛ in. 10% Carbowax 20M 60/80 Chromosorb P column, injection port 220°, column 110°, detector 300°, flow rate 80 cc/min. The retention time of bicyclo[3.2.0]hept-6-en-one under these conditions is 3 minutes 40 seconds. The solvent is removed on a rotary evaporator, leaving a residue of 0.67 g containing some acetone. The ir and nmr spectra of the crude residue are identical with that of bicyclo[3.2.0]hept-6-en-3-one prepared by an alternative method.[6] The product can be further purified by distillation at reduced pressure through a short path apparatus, b.p. 26–28° (3 mm), 49–51° (7 mm), to give 0.34 g (68%) of bicyclo[3.2.0]hept-6-en-3-one.

2. Notes

1. Tropone was prepared by the oxidation of 1,3,5-cycloheptatriene with selenium dioxide as described by Radlick.[7]
2. The photoconversion of 3,5-cycloheptadienone to bicyclo[3.2.0]hept-6-en-3-one can be efficently sensitized using 2-acetonaphthone in ethyl ether, as determined in small-scale runs. The reaction has not been carried out on a preparative scale.

3. Methods of Preparation

Bicyclo[3.2.0]hept-6-en-3-one has been prepared[6] from tropone by hydride reduction to 3,5-cycloheptadienol, photolysis to give epimeric bicyclo[3.2.0]hept-6-en-3-ols, and oxidation to the ketone by chromic acid.

References

1. J. Meinwald, S. L. Emerman, N. C. Yang, and G. Buchi, *J. Amer. Chem. Soc.*, **77**, 4401 (1965).
2. W. E. Parham, R. W. Soeder, and R. M. Dodson, *J. Amer. Chem. Soc.*, **84**, 1755 (1962).
3. W. E. Parham, R. W. Soeder, J. R. Throckmorton, K. Kuncl, and R. M. Dodson, *J. Amer. Chem. Soc.*, **87**, 321 (1965).
4. O. L. Chapman, D. J. Pasto, and A. A. Griswold, *J. Amer. Chem. Soc.*, **84**, 1213 (1962).
5. D. I. Schuster, B. R. Sckolnick, and F. -T. H. Lee, *J. Amer. Chem. Soc.*, **90**, 1300 (1968).
6. O. L. Chapman, D. J. Pasto, G. W. Borden, and A. A. Griswold, *J. Amer. Chem. Soc.*, **84**, 1220 (1962).
7. P. Radlick, *J. Org. Chem.*, **29**, 960 (1964).

BICYCLO[2.1.1]HEXANE

Submitted by R. SRINIVASAN*
Checked by J. PICONE*

1. Procedure

This reaction is carried out in the vapor phase at 1 atm pressure at the boiling point (56°) of 1,5-hexadiene.[1] A cylindrical quartz tube, 8 cm O.D. and 60 cm long with a male 24/40 joint at one end and a female 24/40 joint at the other, is required (Note 1). The quartz tube is mounted vertically along the axis (Note 2) of a Rayonet RPR-208 or RPR-204 reactor fitted with 253.7-nm lamps (Note 3). The bottom end of the tube which should have the male joint is attached to a 100-ml, round-bottomed flask containing 1,5-hexadiene (50 g, 0.6 mole) (Note 4), mercury (5 g) and a few boiling chips. The top of the quartz tube is connected to a water-cooled reflux condenser.

The round-bottomed flask is heated electrically until the 1,5-hexadiene is seen to reflux from the condenser at the top of the quartz tube. The lamps are switched on and the electrical heating is adjusted so that the diene continues to reflux gently (Note 5). The quartz tube should be cleaned after every 48 hours of irradiation in order to remove the polymer film that coats its inside. The progress of the reaction should be followed by vpc. On a 2-meter UCON-550x column at 40°, the retention times of 1,5-hexadiene and bicyclo[2.1.1]hexane are 9.6 and 14.9 minutes, respectively. The other important products and their retention times are allyl cyclopropane, 13.2, and bicyclo[2.2.0]hexane, 23.0 minutes. There are two other minor (<1.0%) products as well. The irradiation should be stopped when the size of the vpc peak due to bicyclo[2.1.1]hexane does not increase in a standard volume of the liquid photolyzate. This time period is typically 5–6 days.

At the end of the irradiation, the liquid photolyzate (32 g) is decanted

* IBM Watson Research Center, Yorktown Heights, New York 10598.

from the mercury and distilled on a 18-in. spinning-band column. The first fraction (3.6 g), b.p. <64°, is made up of 1,5-hexadiene and allyl cyclopropane. The second fraction (9.1 g), b.p. 64.0–74.0°, contains allyl cyclopropane and bicyclo[2.1.1]hexane. The third fraction (1.8 g), b.p. 74.0–84.0°, contains bicyclo[2.1.1]hexane and bicyclo[2.2.0]hexane (Note 6). The residue (16 g) is discarded.

The second fraction is treated four times with a five-fold excess of 30% (weight to volume) aqueous solution of silver nitrate. The aqueous layer is discarded after each treatment. The organic layer is finally washed once with water, dried over calcium chloride, and distilled on a short Vigreux column. Bicyclo[2.1.1]hexane (5.1 g, 10%) of 95% purity distills over. Further purification can be achieved by gas chromatography.

The nmr spectrum of bicyclo[2.1.1]hexane consists of three absorptions at τ 7.47, 8.41, and 9.13 in the ratio 1:3:1. A sample that is 99.7% pure has m.p. 26.5°.

2. Notes

1. This item is commercially available from Southern New England Ultraviolet Co., Middletown, Connecticut 06457, as item No. RQV-324.

2. The quartz tube should be so positioned that the wide portion receives the ultraviolet light to maximum advantage.

3. Available from Southern New England Ultraviolet Co., Middletown, Connecticut 06457. Only mercury resonance lamps can be used in this reaction.

4. Any commercial sample of 1,5-hexadiene of purity greater than 95% is suitable. The sample should be distilled before use.

5. The principal reason for the somewhat poor yield in this reaction is the evaporative loss of 1,5-hexadiene. It is essential that this be kept to a minimum by keeping the material barely at reflux.

6. Gas chromatographic separation of fraction 3 can yield an additional 0.8–1.0 g of bicyclo[2.1.1]hexane.

3. Methods of Preparation

Bicyclo[2.1.1]hexane has been synthesized from norcamphor by photodecarbonylation,[2] from diazonorcamphor by a five-step synthesis,[3] and from 1,5-hexadiene by mercury photosensitization.[4] The last method is the one described here. It is a one-step reaction and uses an inexpensive starting material.

References

1. R. Srinivasan, *J. Phys. Chem.*, **67**, 1367 (1963).
2. R. Srinivasan, *J. Phys. Chem.*, **83**, 2590 (1961).
3. K. B. Wiberg, B. R. Lowry, and T. H. Colby, *J. Phys. Chem.*, **83**, 3998 (1961).
4. R. Srinivasan and F. I. Sonntag, *J. Phys. Chem.*, **89**, 407 (1967).

BICYCLO[2.1.1]HEXAN-2-ONE

Submitted by F. T. BOND, H. L. JONES, and L. SCERBO*
Checked by F. I. SONNTAG and R. SRINIVASAN†

1. Procedure

A. 1,5-Hexadien-3-one. A solution of 76.4 g (0.78 mole) of 1,5-hexadien-3-ol (Note 1) in 4 liters of acetone (Note 2), is cooled to approximately 10°. To this vigorously stirred solution is then added 220 ml of 2.6 M Jones reagent (Note 3) at such a rate that the temperature is maintained below 15°. The addition requires approximately 1 hour, after which the solution is stirred without cooling for an additional hour. At this time 5.0 g of sodium bisulfite is added (Note 4), followed by sufficient solid sodium bicarbonate to neutralize the thick syrup.

The green solution is decanted and the solid residue stirred vigorously with three 300-ml portions of pentane which are combined with the original upper layer. The lower layer which separates is removed (Note 5) and extracted with pentane, then the combined organic layer is washed successively with saturated salt solution. The pentane solution is dried over anhydrous magnesium sulfate. For the preparation of bicyclo[2.1.1]hexan-2-one these solutions can be used directly (Note 6). If desired, the unstable, lachrymatory 1,5-hexadien-3-one can be isolated (Note 7).

B. Bicyclo[2.1.1]hexan-2-one. The irradiations are carried out in a 3.5-liter cylindrical vessel equipped with an efficient condenser, and into

---

* Chemistry Department, University of California, La Jolla, California 92037.
† IBM Research Center, Yorktown Heights, New York 10598.

which is inserted a water-cooled, quartz immersion well (Hanovia No. 19434). The light source is a 450-watt Hanovia lamp (679 A-36).

The solution of crude 1,5-hexadien-3-one is added to the flask and diluted to volume with either additional pentane or acetone (Note 8). After flushing with nitrogen, the solution is stirred magnetically during the course of irradiation. At intervals of 4–6 hours the probe is scraped clean of polymer (Note 9). The course of the irradiation is easily followed by gas chromatography (Note 10); it usually requires about 24–30 hours. The solution is filtered and the pentane removed by distillation through a small column packed with glass helices. The residue is then carefully distilled through a 24-in. spinning-band column to afford a typical yield of 24.1 g (32% from 1,5-hexadien-3-ol), b.p. 56–58° (20 mm) (Note 11). Alternatively, the crude product is converted into its solid sodium bisulfite addition product (Note 12) which is washed thoroughly with cold ether and reconverted into the ketone with 10% sodium hydroxide solution. The basic solution is extracted six times with ether, the combined layers dried over magnesium sulfate, filtered, and the ether removed through a short column, followed by a short path distillation to afford pure ketone. The product is easily identified by its characteristic 1764 cm^{-1} infrared spectrum band (taken on a 2% carbon tetrachloride solution).

2. Notes

1. Prepared by the method of Dreyfuss.[1] It is important that the alcohol be carefully purified before oxidation. Material boiling at 54–56° (20 mm) was commonly used. The checkers used sample supplied by Aldrich Chemical Co. Inc., Milwaukee, Wisconsin 53233.

2. Distilled from potassium permanganate. The checkers used reagent grade acetone without any treatment.

3. Prepared by dissolving 260.0 g of chromium trioxide in 230 ml of concentrated sulfuric acid and diluting to 1 liter with distilled water.

4. This was usually a precautionary step since the green solution indicated complete utilization of the oxidizing agent.

5. At this point and in later washes, difficult emulsions were often encountered. These were broken, as well as possible, through judicial use of salt solution or pentane or both. Alternatively, the crude product could be poured onto saturated salt solution-ice, the pH immediately brought to 6–7 with sodium bicarbonate, and the resulting solution continuously extracted with ether.

6. The yield in the oxidation step is unpredictably variable, and can be estimated by glc if necessary. Standard analysis conditions consisted of a

10-ft DEGS column maintained at 100°. Approximate retention time for 1,5-hexadien-3-one was 10.7 minutes, while the alcohol was eluted at 11.9 minutes. The average yield was 35–40%, although yields as high as 55% have been obtained. In spite of the use of excess oxidizing agent, recovered alcohol is the main impurity. Use of additional Jones reagent results in a lower yield. Many alternative oxidizing conditions (Sarett, manganese dioxide, Oppenauer, dicyclohexylcarbodiimide-DMSO) proved even less successful.

7. The carefully dried pentane solution is filtered and concentrated to about 100 ml on a rotary evaporator at 75–100 mm pressure. The solution is then distilled under vacuum, keeping the pot temperature below 70°, to give 1,5-hexadien-3-one, b.p. 34–40° (30 mm).[2]

8. The irradiation actually proceeds about twice as fast in methanol, although pentane is the more convenient solvent, since overirradiation is not a factor. In general, however, solvent effects on the cyclization appear to be small.[3] The checkers noted that any attempt to conduct the photochemical reaction in a more concentrated solution drastically lowered the yield of bicyclo[2.1.1]hexan-2-one.

9. The lachrymatory nature of the 1,5-hexadien-3-one makes this an unpleasant task unless the irradiation is carried out in a well-ventilated hood.

10. Under the conditions given in Note 6, the photoketone has a retention time of 16.7 minutes.

11. Even under the most careful conditions, traces of 1,5-hexadien-3-ol remain. If this impurity will impair further work, the bisulfite addition compound workup is recommended.

12. Prepared according to the procedure of Vogel.[4]

3. Methods of Preparation

Bicyclo[2.1.1]hexan-2-one has also been prepared by permanganate periodate cleavage of 2-methylenebicyclo[2.1.1]hexane.[5] The present method is essentially that communicated previously.[6]

4. Merits of the Preparation

The procedure offers an easy route into the bicyclo[2.1.1]hexane system.[7] Bicyclo[2.1.1]hexan-2-one has been converted into many other interesting bicyclic compounds including bicyclo[1.1.1]pentane[8] and bicyclo[2.1.1]hex-2-ene.[9] The route described has been adapted to the synthesis of a limited number of substituted bicyclo[2.1.1]hexan-2-ones.[3] Cyclization is not successful with 1,6-heptadien-2-one,[3] only polymer being formed.

References

1. M. P. Dreyfuss, *J. Org. Chem.*, **28**, 3269 (1963).
2. The nmr spectrum of 1,5-hexadien-3-one prepared in this manner is identical with that reported by A. Viola and E. J. Joris, *J. Org. Chem.*, **35**, 856 (1970), ruling out the suggestion by these authors that the preparation gives isomeric 1,4-hexadien-3-one.
3. F. T. Bond, unpublished observations; T. W. Gibson and W. F. Erman, *Abs. Amer. Chem. Soc. Meeting,* Chicago, September 1970; W. Roth and A. Friedrich, *Tetrahedron Lett.* 2607 (1969).
4. A. I. Vogel, *Elementary Practical Organic Chemistry, Part 2,* John Wiley and Sons, New York, 1966, p. 116.
5. J. L. Charlton, P. de Mayo, and L. Skattebøl, *Tetrahedron Lett.*, 4679 (1965).
6. F. T. Bond, H. L. Jones, and L. Scerbo, *Tetrahedron Lett.*, 4685 (1965).
7. For an excellent review see J. Meinwald and Y. C. Meinwald, *Advan. Alicycl. Chem.*, **1**, 1 (1966).
8. J. Meinwald, W. Szkrybalo, and D. R. Dimmel, *Tetrahedron Lett.*, 731 (1967).
9. Meinwald and F. Uno, *J. Amer. Chem. Soc.*, **90**, 800 (1968); F. T. Bond and L. Scerbo, *Tetrahedron Lett.*, 2789 (1968).

N-n-BUTYLPYRROLIDINE

$$CH_3CH_2CH_2CH_2-\underset{H}{N}-CH_2CH_2CH_2CH_3 \xrightarrow{NaOCl} CH_3CH_2CH_2CH_2-\underset{Cl}{N}-CH_2CH_2CH_2CH_3$$

$$\Big\downarrow H_2SO_4$$

$$CH_3CH_2CH_2CH_2-\underset{Cl}{\overset{H}{\underset{|}{N^{\pm}}}}CH_2CH_2CH_2CH_3$$

[pyrrolidine with N-CH₂CH₂CH₂CH₃] $\xleftarrow[(2).\ NaOH]{(1).\ h\nu}$

Submitted by P. G. GASSMAN and R. L. CRYBERG*
Checked by R. D. MILLER†

1. Procedure

Di-*n*-butylamine (12.9 g, 0.10 mole) is added in one portion with stirring to 145 ml of 0.8 *M* (0.116 mole) sodium hypochlorite solution (Note

* The Ohio State University, Columbus, Ohio 43210.
† IBM Research Center, Yorktown Heights, New York 10598.

1) which has been cooled to 0° in an ice-salt bath. After stirring for 15 minutes, the solution is extracted with three 20-ml portions of mixed hexanes (Note 2). The organic extracts are combined, dried over anhydrous magnesium sulfate, and filtered to remove the drying agent. The organic solution is then added dropwise with stirring to 50 ml of 80% sulfuric acid (Note 3) while maintaining the temperature of the sulfuric acid solution below 10° by cooling in an ice-bath. The hydrocarbon phase is separated and extracted with 50 ml of cold 80% sulfuric acid. The combined sulfuric acid solutions are placed in a 30 mm by 300 mm Pyrex test tube and irradiated with a bank of ten 15-watt Sylvania "blacklite" fluorescent lamps for 4 hours (Note 4). When the photolysis is finished (Note 5), the irradiated solution is poured over 1250 g of ice and made strongly basic (pH > 10) by the careful addition of cold 50% aqueous sodium hydroxide. The temperature at the end of the neutralization is generally between 10 and 20°. The resulting basic solution is stirred at room temperature for 3 hours in a stoppered flask. Benzenesulfonyl chloride (15 g) is added and the mixture is stirred for an additional 30 minutes (Note 6). The resulting reaction mixture is extracted with four 25-ml portions of ether. The ethereal extracts are combined and the tertiary amine is extracted from the organic phase with two 25-ml portions of 1:10 hydrochloric acid (Note 7). The acidic solutions are combined, made strongly basic with cold 50% sodium hydroxide solution, and extracted with four 25-ml portions of ether. These ethereal extracts are combined, dried over anhydrous magnesium sulfate, and the drying agent removed by filtration. The ether is removed by distillation through an 8-in. column packed with glass helices. The residue is fractionally distilled to give 7.3–9.9 g (58–75%, Note 8) of *N-n*-butylpyrrolidine, b.p. 83–86° (75 mm), n^{25}D 1.4355.

2. Notes

1. Commercial household bleach (Clorox, Purex, etc.) is a satisfactory sodium hypochlorite source. The molarity of the commercial bleach is readily determined by titration.[1]
2. Skelly Solvent B is used. Before use it is washed free of olefins by stirring with concentrated sulfuric acid for 24 hours.
3. The 80% sulfuric acid is prepared by the addition of 100 g of concentrated sulfuric acid to 15 g of water.
4. The ten fluorescent tubes are arranged around the inside of a 12.75-in. diameter stainless steel cylinder and the sample is situated in the center of the circular bank of lights. A commercial apparatus of this type is available from Southern New England Ultraviolet Company.

5. The progress of the reaction is followed by removing one or two drops of the solution being irradiated, diluting with 2 ml of water, and adding a crystal of sodium iodide. The formation of a yellow color indicates the presence of active chlorine and shows that the photolysis is incomplete.

6. The benzenesulfonyl chloride treatment is used to remove primary and secondary amines which may be present in small amounts.

7. The acid extraction is used to separate the desired tertiary amine from the small amounts of benzenesulfonamides formed in the preceding step.

8. The yields are somewhat variable. The average yield for six runs is 63%.

3. Methods of Preparation

The procedure described is a modification of that described in the literature.[2,3] N-n-Butylpyrrolidine has also been prepared *via* thermal decomposition of protonated di-n-butyl-N-chloramine.[4]

4. Merits of the Preparation

The procedure outlined here provides a general description of the photochemical Hofmann-Löffler-Freytag reaction. A variety of tertiary amines have been prepared by this method.[5]

References

1. I. M. Kolthoff and E. B. Sandell, *Textbook of Quantitative Inorganic Analysis*, 3rd ed., The Macmillan Company, New York, 1952, p. 597.
2. P. G. Gassman and D. C. Heckert, *Tetrahedron*, **21**, 2725 (1965).
3. R. S. Neale and M. R. Walsh, *J. Amer. Chem. Soc.*, **87**, 1255 (1965).
4. G. H. Coleman, G. Nichols, and T. F. Martens, *Org. Syn. Coll. Vol.*, **3**, 159 (1955).
5. For a review see M. Wolff, *Chem. Rev.*, **63**, 55 (1963).

CYCLOBUTENE

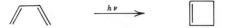

Submitted by F. I. SONNTAG and R. SRINIVASAN*
Checked by J. PICONE*

1. Procedure

The apparatus consists of a cylindrical irradiation vessel made of quartz, 40 cm long and 5 cm O.D., to one end of which is attached a narrow piece of quartz tubing, 5 cm long and 8 mm O.D., which can be closed hermetically by a small serum cap (Note 1). The source of ultraviolet (2537 A) radiation is a Rayonet Reactor (Note 2).

The container is charged with 500 g cyclohexane and cooled in ice water. At the onset of crystallization the cooling is discontinued, and 1,3-butadiene (Note 3) is bubbled into the solvent until 78.0 g (1.4 moles) has been absorbed. The container is closed with a serum cap (Note 4) and suspended in the irradiation cavity. Irradiation is continued for 1200 hours. During this period the solution is transferred to a clean container after 150, 400, and 800 hours (Notes 5, 6). At the end of this period (Note 7) the irradiated solution is cooled in Dry Ice-acetone to about −20°, poured into a 1-liter flask, precooled to the same temperature, and submitted to distillation on a 22-in. spinning-band column (Note 8). The mixture is heated slowly (during 30 minutes) to boiling and kept at this temperature for about 1 hour. The column is operated under total reflux; the C_4-hydrocarbon fraction, which does not condense under these conditions, is allowed to pass through the reflux condenser and is collected in a 500-ml, round-bottomed flask immersed in Dry Ice-acetone. In order to collect any uncondensed C_4-hydrocarbons the receiving flask is followed by two small (20-ml and 5-ml) traps immersed in the same cooling bath. When the distillation is completed, small quantities of benzene are added to the cold traps, the benzene is allowed to melt over the distillate, and the solution thus obtained is added to the bulk of the distillate in the receiver. Finely ground maleic anhydride (50 g, 0.5 mole) and 150 ml of benzene are added to the distillate while this is being kept in a Dry Ice-acetone bath at approximately −20°. After the receiver has been closed by a rubber stopper (and kept tight by a clamp), the cooling bath is removed and the

* IBM Research Center, Yorktown Heights, New York 10598.

contents are stirred magnetically, while allowed to warm up to room temperature, until the maleic anhydride dissolves. The clear solution is left standing for 48 hours at room temperature. Turbidity appears after 1 day and a white precipitate eventually settles at the bottom. The reaction mixture is cooled to about $-20°$ before removing the stopper and submitted to distillation on the spinning-band column, as described before. The distillate consists of 24.0–30.0 g of cyclobutene (yield 31–38%). The infrared spectrum of the product[1] (taken on a gas sample at 100 torr, 10 cm path length) does not show absorptions attributable to butadiene[2] or bicyclobutane.[3]

2. Notes

1. Two such containers are required.
2. Model RPR-208 (supplied by Southern New England Ultraviolet Company, Middletown, Connecticut) fitted with RUL-2537 lamps. The quartz vessel is placed in the center and cooled by a centrifugal air blower mounted on the base of the reactor to blow air upward.
3. Supplied by Matheson Gas Products, East Rutherford, New Jersey 07073. It is essential that the sample designated "research grade" and stated to be 99.7 mole % pure be used. Material of lesser purity does not give satisfactory yields.
4. The rapid deterioration of the serum cap appears to be due to the direct action of the ultraviolet radiation rather than that of ozone. Wrapping the serum cap with a piece of aluminum foil gives adequate protection.
5. The necessity of transferring the solution to a clean container is dictated by the increasing opacity of the quartz vessel to ultraviolet light due to the formation of polymeric material.

A considerable loss of volatile C_4-hydrocarbons is experienced when the transfer is effected by pouring the solution into the clean vessel, in spite of preliminary cooling. In order to eliminate these losses the following method is found to be satisfactory: the duplicate quartz container, closed with a serum cap, is thoroughly evacuated by connecting to an oil vacuum pump *via* a 22 gauge hypodermic needle. The cooled vessel containing the irradiated solution is connected with the evacuated container by a two-way hypodermic needle (made by cutting off the wide end of a normal hypodermic needle), care being taken that both serum caps be punctured simultaneously. The pressure differential ensures a fast transfer of the solution. Cooling of the receiving vessel and gentle warming of the donating vessel will accelerate the process.

6. The polymer is removed from the used container either with chromic

acid cleaning mixture or by heating the quartz tube in a hot flame in the presence of a stream of air.

7. On the basis of the original amount of butadiene, a conversion of about 50% to cyclobutene has taken place at this point, while less than 20% of butadiene is left unchanged.

8. The use of a spinning-band column is not essential as, owing to the great difference in boiling points (cyclobutene 2°, butadiene —4°, cyclohexane 81°), no fractionation is necessary.

3. Methods of Preparation

Cyclobutene has been prepared by the Hofmann elimination method from cyclobutyltrimethylammonium hydroxide,[4-6] the decomposition of cyclobutyldimethylamine oxide,[7] and the pyrolysis of the adduct prepared from 1,3,5-cyclooctatriene and dimethyl acetylenedicarboxylate.[8]

4. Merits of the Preparation

The present procedure has the advantage of using inexpensive, commercially available starting materials. The few steps involved require relatively simple handling. It can be adapted to the synthesis of a variety of alkyl-substituted cyclobutenes.

References

1. R. C. Lord and D. G. Rea, *J. Amer. Chem. Soc.*, **79**, 2401 (1957).
2. American Petroleum Institute Research Project 44, Ultraviolet Spectral Data, Serial No. 65.
3. I. Haller and R. Srinivasan, *J. Chem. Phys.*, **41**, 2745 (1964).
4. Willstätter and W. von Schmaedel, *Ber.*, **38**, 1992 (1905).
5. R. Willstätter and J. Bruce, *Ber.*, **40**, 3979 (1907).
6. G. B. Heisig, *J. Amer. Chem. Soc.*, **63**, 1698 (1941).
7. J. D. Roberts and C. W. Sauer, *J. Amer. Chem. Soc.*, **71**, 3925 (1949).
8. A. C. Cope, A. C. Haven, Jr., F. L. Ramp, and E. R. Trumbull, *J. Amer. Chem. Soc.*, **74**, 4867 (1952).

1β,5-CYCLO-5β,10α-CHOLESTAN-2-ONE
(LUMICHOLESTENONE)

Submitted by B. A. SHOULDERS, W. W. KWIE, and P. D. GARDNER*
Checked by T. S. CANTRELL†

1. Procedure

A. Using commercial light sources. A solution of 10.0 g (0.026 mole) of twice recrystallized 4-cholesten-3-one (Note 1) dissolved in 4.0 liters of *t*-butyl alcohol (Notes 2, 3) is placed in a 5-liter reactor flask outfitted with a fritted dip tube for admission of nitrogen, a port for pipet sampling, and a centrally located, water-cooled immersion well for placement of a lamp (Note 4). The sampling port is fitted with an easily removed adapter which is connected by rubber tubing to a mercury gas valve. Nitrogen (commercial prepurified) is admitted through the dip tube at a fairly rapid rate for 2 hours to remove most of the oxygen from the system. Without diminishing the rate of gas flow, the adapter is removed for withdrawal of a 1-ml sample for monitoring purposes and then replaced. After approximately 5 minutes, the gas flow is reduced to a very slow rate and the lamp is turned on. The 1-ml aliquot is suitably diluted with 95% ethanol and its optical density determined at 240 mμ as a measure of the $t = 0$ concentration of substrate. Periodic withdrawal and optical reading of such samples during the irradiation thus provides data on the rate of disappearance of substrate (Note 5). Irradiation is discontinued when the intensity of the 240 mμ band drops by 75–85%. The resulting suspension is concentrated using aspirator vacuum to about 200 ml and then filtered with suction to remove the small amount (0.1–0.5 g) of very slightly soluble photodimer. The filtrate is concentrated to dryness at an aspirator with the aid of a steam bath. The viscous residue is then dissolved in 30–40 ml. of 95% ethanol and cooled at 0° until crystallization occurs. The yield of crude product

* University of Utah, Salt Lake City, Utah 84112.
† Rice University, Houston, Texas 77001.

obtained by suction filtration is 2.0–2.5 g and has a melting range of 160–164° (Note 6). One recrystallization from ethanol gives 1.6–2.2 g of essentially pure lumicholestenone, m.p. 164–166°. Chromatographically pure material has m.p. 165–166°. An additional 0.3–0.6 g of product can be obtained by concentrating the combined crystallization mother liquors to dryness and chromatographing the residue on neutral alumina using 5% ethyl ether in pentane or hexane. Lumicholestenone appears in the first few fractions. The contributors feel that this small addition to the yield does not warrant the labor of a chromatographic recovery.

B. *Using sunlight.* Two Pyrex tubes 10 cm in diameter and 200 cm long are sealed off at one end and necked down to approximately 2-cm length of 2-cm diameter tubing such that this end can be sealed off with a torch after the tubes have been filled. Each tube is filled with a solution of 45 g of 4-cholesten-3-one dissolved in 13 liters of *t*-butyl alcohol. Each is then connected to an aspirator by a piece of rubber tubing containing an in-line stopcock for control of vacuum and cautiously evacuated so as to maintain a vigorous boiling of the solution. When approximately 100 ml of solvent has evaporated, the tube is sealed off with a torch. The tubes are placed in a nearly horizontal position in direct sunlight during July or August and left for 5 weeks. Processing of the irradiated solution as described in Section A gives 5–6 g of photodimer. Chromatography of the remaining residue on 8 lb of neutral alumina gives 19–22 g of pure lumicholestenone.

2. Notes

1. See reference 2. Recrystallization from acetone-methanol (1:1) gives satisfactory material.
2. *t*-Butyl alcohol, m.p. 24–25.5°, from Eastman Organic Chemicals is quite satisfactory without purification. Solvent which has been recovered from a previous preparation is also suitable.
3. The use of dilute solutions helps to suppress the competitive photodimerization of 4-cholesten-3-one.
4. The immersion well (thimble) should be constructed of Pyrex or, if a quartz thimble is used, a Pyrex liner is advisable to minimize the buildup of an insoluble film on the surface of the thimble by absorbing very short wavelength irradiation. Except for low-pressure mercury resonance (2537 Å) lamps, virtually any ultraviolet source which will fit into the thimble can be used. The contributors used Hanovia 200-watt and 450-watt lamps routinely. Reaction time can be reduced considerably with more intense sources such as the General Electric AH-6 high-pressure unit. This

lamp requires a specially constructed U-tube of sufficiently compact design to fit into the thimble. It was operated routinely in a vertical position, although the manufacturer cautions that it should be operated only horizontally.

5. This reaction is photochemically very inefficient, requiring approximately 170 hours with a 200-watt lamp, 75 hours with a 450-watt lamp, and 24 hours with a 1000-watt lamp for 80% conversion of starting material.

6. Only infrequently does the product fail to crystallize at this stage. Then the residue is dissolved in the minimum volume of 1:1 benzene-ligroin and charged to a column of 500 g of neutral alumina. Elution with 5% ether in petroleum ether (30–60°) brings off the product in the first few fractions. The yield is about the same as that obtained by direct crystallization.

References

1. W. W. Kwie, B. A. Shoulders and P. D. Gardner, *J. Amer. Chem. Soc.*, **84**, 2268 (1962); B. A. Shoulders, W. W. Kwie, W. Klyne, and P. D. Gardner, *Tetrahedron*, **21**, 2973 (1965).
2. *Orig. Syn. Coll. Vol.*, **3**, 207 (1955).

cis, trans-1,3-CYCLOOCTADIENE

Submitted by P. G. GASSMAN and E. A. WILLIAMS*
Checked by T. D. ROBERTS and C. GAMBILL†

1. Procedure

cis,cis-1,3-Cyclooctadiene (60 g, 0.56 mole) (Note 1), 2 g of acetophenone, and 500 ml of mixed pentanes (Note 2) are placed in a 700 ml irradiation vessel equipped with a water-cooled internal Pyrex immersion well containing a 450-watt Hanovia medium-pressure lamp, provisions for

* The Ohio State University, Columbus, Ohio 43210
† University of Arkansas, Fayetteville, Arkansas 72701.

efficient magnetic stirring (Note 3) and for maintaining a slow stream of nitrogen through the solution. The solution is irradiated under nitrogen with stirring for 4 hours while maintaining the solution temperature between 10 and 15° with an ice bath. At this time the irradiation is stopped and the outer wall of the immersion well is washed with acetone to remove any polymer that has formed. The irradiation is continued for an additional 4 hours under the same conditions, and the washing procedure repeated. The irradiation is continued for another 4 hours, then stopped, and the solution transferred to a 2-liter, three-necked, round-bottomed flask equipped with a mechanical stirrer. The solution is cooled to 0° in an ice bath, and 200 ml of 20% silver nitrate solution is added with stirring. Vigorous stirring is continued for 2 hours while maintaining the temperature at 0°. The solution is filtered and the precipitate washed with water and several portions of mixed pentanes. The product is permitted to dry thoroughly on the filter paper overnight to yield 30–34 g of the silver nitrate complex of *cis,trans*-1,3-cyclooctadiene (*ca.* 80% based on reacted *cis,cis*-1,3-cyclooctadiene) (Note 4), m.p. 126.5–127.5° (dec.). For very pure material the complex is recrystallized from methanol to yield 22–26 g of complex (75% recovery), m.p. 127.5–128° (dec.) (Note 5).

To a stirred solution of 40 g of complex (0.14 mole) in 160 ml of pentane is added dropwise 40 ml of concentrated ammonium hydroxide over a 30-minute period. The layers are separated and the aqueous layer is extracted with two 100-ml portions of pentane. The combined organic layers are washed with two 100-ml portions of water, one 100-ml portion of saturated sodium chloride solution, dried over anhydrous magnesium sulfate, filtered, and the solvent removed by distillation at reduced pressure. The *cis,trans*-1,3-cyclooctadiene is flash-distilled to yield 9.0–10.5 g of product (60–70%), b.p. 25° (0.1 mm).

2. Notes

1. *cis,cis*-1,3-Cyclooctadiene obtained from Columbian Carbon Co. was used without purification.
2. Skelly Solvent F was used.
3. Efficient stirring is necessary to effect thorough mixing of the solution and good conversion into *cis,trans*-1,3-cyclooctadiene.
4. On a larger scale, the yield based on reacted *cis,cis*-1,3-cyclooctadiene is approximately the same, but the conversion into the *cis,trans*-1,3-cyclooctadiene is lower.
5. Use of the unrecrystallized complex in the generation of the *cis,trans*-1,3-cyclooctadiene results in pure material.

3. Methods of Preparation

The procedure described is a modification of that described in the literature.[1,2]

4. Merits of the Preparation

This procedure describes a practical method for the generation of *cis,-trans*-1,3-cyclooctadiene.

References

1. R. S. H. Liu, *J. Amer. Chem. Soc.*, **89**, 112 (1967).
2. W. J. Nebe and G. J. Fonken, *J. Amer. Chem. Soc.*, **91**, 1249 (1969).

Dimer of 2-Aminopyridine

Submitted by EDWARD C. TAYLOR and GAVIN G. SPENCE*
Checked by B. R. VOGT and R. P. SCELIA†

1. Procedure

The source of ultraviolet (2537 Å) radiation is a large Hanau low-pressure immersion lamp. The filament of the lamp is 78 cm long and is enclosed in a quartz tube 109 cm long and 4.5 cm O.D. The reaction vessel is a Pyrex tube, 91.5 cm long and 5.1 cm O.D., designed to fit snugly around the lamp (Notes 1, 2). The reaction vessel is cooled by a water jacket 95 cm long and 5.1 cm O.D. fitted with a water inlet at the bottom and two outlets at the top. The water stream is precooled by passing it through a coil of copper tubing immersed in an ice bath.

The reaction vessel is filled with a 150-ml solution of 78.0 g (0.829 mole) of 2-aminopyridine (Note 3) in concentrated hydrochloric acid.

---

* Princeton University, Princeton, New Jersey 08540.
† Union Carbide Research Center, Eastview, New York 10591.

The solution is irradiated, and the product, which separates from the solution as white crystals, is harvested regularly (Note 4). After irradiation for 40 hours (Note 5), the yield is 59.5 g (55.0%) of white crystals (m.p. 215–218°, dec., rapid heating) (Notes 6, 7).

2. Notes

1. Since the reaction vessel is only slightly larger in diameter than the lamp, a relatively small volume (150 ml) completely fills the vessel. The solution is held in a thin layer around the lamp, and the radiation from the lamp is thus used very efficiently.

2. The checkers used a Hanovia 450-watt high-pressure mercury vapor lamp with a quartz window 110 mm long and 19 mm O.D. The lamp was positioned inside a water-cooled Pyrex immersion well having a 50-mm O.D. and extending 220 mm below a 60/50 standard taper joint. The solution was contained in a glass jacket (surrounding the well) having a 64 mm O.D. and extending 240 mm below the 60/50 standard taper joint.

3. 2-Aminopyridine is available from Matheson, Coleman & Bell (cat. No. AX1155). It is best purified by distillation at atmospheric pressure.

4. As the irradiation proceeds, solid collects in the reaction vessel and interferes with further absorption of light by the solution. Thus it is necessary to collect the product periodically. The solid is removed by filtration and washed with ethanol. The solid adhering to the lamp and to the reaction vessel is removed by washing it into a beaker with ethanol.

5. The reaction becomes increasingly inefficient as the concentration of 2-aminopyridine is decreased by removal as the dimer. Thus 19.7 g is collected after irradiation for 8 hours, but following 8-hour irradiation periods yield 13.7, 14.0, 8.3, and 3.7 g, respectively. Further irradiation is unnecessary because the reaction is very inefficient at lower concentrations.

6. The melting point of the dimer is dependent on the rate of heating. It is not a true melting point, for at high temperatures the dihydrochloride of the dimer breaks down to the hydrochloride of 2-aminopyridine (m.p. 86°) which then melts immediately. The lower the rate of heating, the lower will be the apparent melting point of the dimer. The reported melting point was obtained on a Thomas-Hoover apparatus with the heat dial set at 8.

7. The product can be recrystallized from *ca.* 200 ml of 40% aqueous ethanol.

3. Methods of Preparation

The only reported synthesis of this dimer utilizes this procedure.[1,2]

4. Merits of the Preparation

The reaction is carried out very simply, and the dimer may be prepared on a relatively large scale. The only workup required is filtration and washing of the product, which separates directly from the irradiated solution.

References

1. E. C. Taylor, R. O. Kan, and W. W. Paudler, *J. Amer. Chem. Soc.*, **83**, 4484 (1961).
2. E. C. Taylor and R. O. Kan, *J. Amer. Chem. Soc.*, **85**, 776 (1963).

5,5-DIMETHYL-1-VINYLBICYCLO[2.1.1]HEXANE

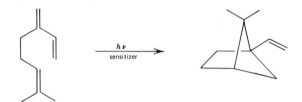

Submitted by R. S. H. LIU*
Checked by J. CHU and A. S. KENDE†

1. Procedure

A solution of 13.6 g (0.1 mole) of freshly distilled myrcene (Note 1) and 0.2 g of Michler's ketone (Note 2) in 500 ml of anhydrous ether (Notes 3, 4) is added to a water-cooled Pyrex immersion apparatus (Note 5) equipped with a reflux condenser and a magnetic stirrer. A stream of nitrogen is passed through the solution for approximately 5 minutes to remove dissolved oxygen. While kept under nitrogen and stirred, the solution is irradiated with a 450-watt Hanovia medium-pressure mercury lamp. After approximately 24 hours the reaction is complete (Note 6).

The solution is transferred to a round-bottomed flask and concentrated by evaporation on a rotary evaporator. The concentrated mixture is distilled under vacuum through a 24-in. spinning-band column. The colorless

---

* Department of Chemistry, University of Hawaii, Honolulu, Hawaii 96822.
† University of Rochester, Rochester, New York 14627.

liquid, which boiled at 56–57° (20 mm) (10.2 g, 75% yield) (Note 7) is essentially pure 5,5-dimethyl-1-vinylbicyclo[2.1.1]hexane, $n^{19.2}$D 1.4641 (Note 8).

2. Notes

1. The commercial sample of myrcene (Aldrich Chemical, technical grade) is only 80–85% pure. Samples of approximately 95% purity can be obtained by careful fractional vacuum distillation of the commercial sample through a 24-in. spinning-band column. Since the impurities do not interfere with the photoreaction, the sample can be satisfactorily used after one distillation.

2. Michler's ketone, 4,4'-bis(dimethylamino)benzophenone (Eastman Kodak, practical grade, twice recrystallized from methanol) is used because of its broad absorption band and its lack of chemical reaction with myrcene. Other compounds such as benzophenone, triphenylene, and 2-acetonaphthone are also acceptable sensitizers.

3. Other inert solvents such as n-pentane and benzene may be used in place of ether.

4. At higher concentrations of myrcene, photosensitized dimerization becomes an important side reaction.

5. The Pyrex glass of the immersion irradiation well serves to filter off the 2537 Å light to avoid direct irradiation processes.

6. Unnecessarily prolonged irradiation should be avoided. The progress of the reaction can be conveniently followed by glpc analyses. On all columns tested, the retention time of the product is much shorter than that of myrcene. One satisfactory condition for analyses is a 6-ft 15% Apiezon J column at 90°.

7. The quantum yield of the reaction, when a high-energy sensitizer such as Michler's ketone or benzophenone is used, is 0.023.[1]

8. The ir and nmr spectra of the product have been reported.[1]

3. Methods of Preparation

The irradiation procedure described here is based on that which has already been published.[1] Similar procedures have been used in the preparation of 2-methylenebicyclo[2.1.1]hexane,[1,2] 5,5-dimethyl-2-methylenebicyclo[2.1.1]hexane, and 1-vinylbicyclo[2.1.1]hexane.[1]

References

1. R. S. H. Liu and G. S. Hammond, *J. Amer. Chem. Soc.*, **89**, 4936 (1967).
2. J. L. Charlton, P. de Mayo and L. Skattebøl, *Tetrahedron Lett.*, 4679 (1965).

4,5-DIPHENYL-1-METHYLIMIDAZOLE

$$\underset{C_6H_5}{\overset{C_6H_5}{\diagdown}}\!\!\!\underset{C=N}{\overset{C=N}{\diagup\diagdown}}\!\!\!\overset{CH_2}{\underset{CH_2}{\diagup}} \xrightarrow{h\nu,\ C_2H_5OH} \underset{C_6H_5}{\overset{C_6H_5}{\diagdown}}\!\!\!\underset{C-N}{\overset{C=N}{\diagup\diagdown}}\!\!\!\overset{}{\underset{CH_3}{\diagup}}\!\!\!C-H$$

Submitted by JOHN L. MIESEL and PETER BEAK*
Checked by R. D. MILLER†

1. Procedure

In a vessel equipped with a nitrogen inlet, reflux condenser, and insert containing a water-cooled quartz envelope, a solution of 1.5 g (0.0064 mole) of 2,3-dihydro-5,6-diphenylpyrazine (Note 1) in 420 ml of absolute ethanol (Note 2) is photolyzed under a stream of dry nitrogen with a Pyrex-filtered Hanovia type L 450-watt high-pressure mercury vapor lamp (Note 3). After 6 hours the reaction is essentially completed. The solvent is removed under reduced pressure, and the crude product is recrystallized from ether to give colorless rectangular crystals, m.p. 159–160°. The yield is in the range 900–1050 mg (60–70%) (Note 4).

2. Notes

1. 2,3-Dihydro-5,6-diphenylpyrazine is available commercially from Aldrich Chemical Company.
2. An anhydrous solvent must be used. Hydrocarbon solvents are also suitable.
3. Corex or Vycor filters or the unfiltered lamp may also be used. A Hanovia 23-watt type SC-2537 low-pressure Vycor mercury-vapor immersion lamp was also successfully used.
4. Higher yields can be obtained by chromatography.

3. Methods of Preparation

4,5-Diphenyl-1-methylimidazole can be prepared by methylation of 4,5-diphenylimidazole with methyl sulfate.[1] General methods of imidazole synthesis have been reviewed.[2]

---

* Department of Chemistry, University of Illinois, Urbana, Illinois 61801.
† IBM Research Center, Yorktown Heights, New York 10598

4. Merits of the Preparation

The photorearrangement of 2,3-dihydro-5,6-diphenylpyrazine to 4,5-diphenyl-1-methylimidazole is typical of a general reaction useful for the preparation of novel and known alkyl and aryl imidazoles. It appears to involve initial ring opening to an enediimine followed by ring closure and proton transfer.[3]

References

1. A. Simonov and A. Garnovskii, *Zh. Obshch. Khim.*, **31**, 114 (1961).
2. E. S. Schipper and A. R. Day, in R. C. Elderfield, *Heterocyclic Compounds*, Vol. 5, John Wiley and Sons, New York, 1953, p. 194.
3. P. Beak and J. L. Miesel, *J. Amer. Chem. Soc.*, **89**, 2357 (1967).

4,4-DIPHENYL-3-OXATRICYCLO[4.2.1.0^{2,5}]NONANE

The oxetane from the photocycloaddition of benzophenone to norbornene

$C_6H_5CC_6H_5$ + [norbornene] $\xrightarrow{h\nu}_{C_6H_6}$ [product]

Submitted by D. R. ARNOLD, A. H. GLICK, and V. Y. ABRAITYS*
Checked by R. D. MILLER†

1. Procedure (Note 1)

In a 250-ml irradiation vessel (Note 2), 6.6 g (0.036 mole) of benzophenone and 3.8 g (0.040 mole) of norbornene are dissolved in 200 ml of benzene. The solution is irradiated through a Pyrex immersion well which is cooled by means of circulating water ($\sim 15°$). The light source is a 450-watt high-pressure mercury vapor lamp (Note 3).

After 12 hours the infrared spectrum of an aliquot indicates the complete disappearance of the carbonyl absorption (6.0 µ) from benzophenone and the appearance of the characteristic oxetane absorption band at 10.2 µ.

---

* Union Carbide Research Institute, P.O. Box 278, Tarrytown, New York 10591.
† IBM Research Center, Yorktown Heights, New York 10598.

The solvent is removed under reduced pressure and the resulting yellow oil (9.97 g) chromatographed on a column of neutral alumina, 150 g (Note 4), eluting with hexane (Note 5).

The product, 8.1 g (81%), is obtained as colorless crystals, m.p. 120–124°. Recrystallization from n-hexane raised the melting point to 124–125° (Note 6).

2. Notes

1. The same procedure can be used in the preparation of a wide variety of oxetanes. For a review of this reaction see Reference 1.

2. The irradiation apparatus is commercially available from Ace Glass, Inc., Vineland, New Jersey (reaction vessel, cat. No. 6522-05; immersion well, cat. No. 6515-25).

3. The irradiation lamp and power supply are available from Ace Glass, Inc., (cat. Nos. 6515-34, 6515-60) and from Engelhard Hanovia, Inc., Newark, New Jersey (cat. Nos. 679A-36, 34245-1[8354-1]).

4. The submitters used reagent grade aluminum oxide obtained from Merck & Co., Rahway, New Jersey. The checker found Woelm alumina (neutral, activity I) unsuitable for this purpose.

5. The major impurity is benzopinacol, which can be isolated by further elution with benzene.

6. Reported melting points are 121°;[2] 128–129°.[3]

References

1. D. R. Arnold, *Advan. Photochem.*, **6**, 301–423 (1968).
2. D. Scharf and F. Korte, *Tetrahedron Lett.*, 821 (1963).
3. D. R. Arnold, R. L. Hinman, and A. H. Glick, *Tetrahedron Lett.*, 1425 (1964).

ETHYL 1-HYDROXYCYCLOHEXANECARBOXYLATE

$$\text{C}_6\text{H}_{11}\text{-CO}_2\text{C}_2\text{H}_5 + \text{CH}_3\text{CO}_3\text{H} \xrightarrow{h\nu} \text{C}_6\text{H}_{10}(\text{OH})\text{-CO}_2\text{C}_2\text{H}_5 + \text{CH}_4\uparrow + \text{CO}_2\uparrow$$

Submitted by D. L. HEYWOOD*
Checked by T. D. ROBERTS†

1. Procedure

The apparatus consists of a 5-liter, three-necked flask fitted with a stirrer in one neck, a water-cooled quartz immersion well containing a 100-watt Hanovia high-pressure mercury arc in the center neck (Note 1), and a reflux condenser vented to the atmosphere in the third. The flask is also fitted with a thermometer port (Note 2).

The flask is charged with 1460 g (9.46 moles) of ethyl cyclohexanecarboxylate (Note 3). While being continuously irradiated, a solution (1226 g) of peracetic acid (304 g, 9.46 moles) in ethyl acetate is added (Note 4). The temperature is maintained at 25–30° with ice-water bath cooling. The addition requires 5 hours, and the solution is allowed to react under irradiation for an additional 43 hours at 50°, at the end of which time the conversion is greater than 95% on the basis of peracetic acid titration (Note 5). The solution is washed with cold dilute sodium bicarbonate, dried, and distilled to give, after recovery of ethyl acetate, 940 g of unreacted ethyl cyclohexanecarboxylate and 442 g of crude ethyl hydroxycyclohexanecarboxylate, b.p. 85–120° (16–3 mm) (Note 6). This crude fraction, which contains an additional 148 g of unconverted starting material and small amounts of other hydroxylated products, is carefully fractionated to give 170 g (25% yield) of pure ethyl 1-hydroxycyclohexanecarboxylate, b.p. 101–103° (15 mm) (Note 7). Saponification equiv.: calc., 172.2; found, 173.5. A sample was saponified in the usual fashion to give 1-hydroxycyclohexanecarboxylic acid, m.p. and mixed m.p. with authentic 1-hydroxycyclohexanecarboxylic acid, 108–109°. Higher-boiling isomeric ethyl hydroxycyclohexanecarboxylates (b.p. 103–125°) are formed in the reaction, but the 1-isomer is the major product.

* Union Carbide Corporation, P.O. Box 8361, South Charleston, West Virginia 25303.
† University of Arkansas, Fayetteville, Arkansas 72701.

2. Notes

1. A variety of other systems such as external radiation and other ultraviolet sources can be used provided that adequate temperature control is maintained, provision for venting gases is present, and quartz or some other material to allow transmission of light of wavelength less than about 2900 Å is used.

2. Cleaner results were obtained when the reaction system (liquid phase included) was swept gently with dry nitrogen before initiation of reaction to purge the oxygen.

3. Cyclohexanecarboxylic acid b.p. 130–133° (20 mm) was esterified with ethanol in the usual manner to give ethyl cyclohexanecarboxylate, b.p. 79–83° (16 mm).

4. Peracetic acid in an "inert" solvent such as ethyl acetate or acetone is preferred; see reference 1.

5. Peracetic acid is determined by any of a variety of the standard peroxide methods via potassium iodide-sodium thiosulfate titration. Visual assessment of the degree of completion of reaction is made by observing the decreasing rate of gas evolution (methane and carbon dioxide) which is evident throughout the reaction.

6. This fraction consists of a mixture of isomers of monohydroxy products at this stage, in addition to small amounts of ketonic products. These are minimized by using as large an excess of hydrocarbon in the initial charge as practicable.

7. The checker found that the purity of the peracetic acid determined the yield.

3. Methods of Preparation

The ethyl ester of 1-hydroxycyclohexanecarboxylic acid is an old compound and has usually been prepared from the acid by standard esterification methods. The acid is classically prepared from the cyanohydrin,[2,3] but has also been prepared by a Favorskii reaction on 2-chlorocycloheptanone[1] and by a Canizzaro reaction on 1-hydroxycyclohexanecarboxaldehyde.[4] An interesting analog to the present method is reported for the methyl ester,[5] in which methylcyclohexanecarboxylate is autoxidized and the resulting 1-hydroperoxide is reduced to the 1-hydroxy ester.

4. Merit of the Preparation

The present procedure would not offer merit for preparing 1-hydroxycyclohexanecarboxylate in comparison with, e.g., a standard cyanohydrin reaction on cyclohexanone. Rather, it is offered as an example of the technique that can be used to afford a great variety of otherwise difficultly accessible alcohols. In consideration of the relatively indiscriminate nature of the reaction, aliphatic hydrocarbons can be oxidized to a spectrum of monohydroxylated products. This technique has been used successfully for the preparation of 2-hydroxyadamantane[6] and a variety of other alcohols.[7]

References

1. B. Phillips, F. C. Frostick, Jr., and P. S. Starcher, *J. Amer. Chem. Soc.*, **79**, 5982 (1957); B. Phillips, P. S. Starcher, and B. D. Ash, *J. Org. Chem.*, **23**, 1823 (1958).
2. Beilstein, 4th ed., Vol. X, p. 4; 1st suppl., p. 4; 2nd suppl., p. 4.
3. J. Rouzaud, G. Cauquil, and L. Giral, *Bull. Soc. Chem. Fr.*, 2908 (1964).
4. E. D. Venus-Danilova and V. F. Kazimirova, *J. Gen. Chem. USSR*, **7**, 2639 (1937) [*Chem. Abstr.*, **32**, 2099 (1938)].
5. J. J. Fuchs, U.S. Patent 3125600 (March 17, 1964), to E. I. du Pont de Nemours & Co.
6. P. von R. Schleyer and R. D. Nicholas, *J. Amer. Chem. Soc.*, **83**, 182 (1961).
7. D. L. Heywood, H. A. Stansbury, Jr., and B. Phillips, U.S. Patent 3182008 (May 4, 1965), to Union Carbide Corp.; *J. Org. Chem.*, **26**, 281 (1961).

3-FLUOROPHENANTHRENE

Submitted by CLELIA W. MALLORY and FRANK B. MALLORY*
Checked by EDWARD C. TAYLOR and BEN E. EVANS†

1. Procedure

The apparatus includes a quartz immersion well with a water cooling jacket; alternatively, the similarly designed Hanovia 19433 Vycor immer-

---

* Bryn Mawr College, Bryn Mawr, Pennsylvania 19010.
† Princeton University, Princeton, New Jersey 08540.

sion well may be used. The light source is a 100-watt General Electric H100 A4/T mercury lamp modified by cutting away the outer glass envelope[1] and operated using a General Electric 9T64Y-3518 transformer; alternatively, the Hanovia 200-watt 654A-36 mercury lamp may be used with the appropriate power supply (Note 1).

A solution of 1.98 g (0.01 mole) of *trans-p*-fluorostilbene (Note 2) and 0.127 g (0.5 mmole) of iodine in 1 liter of cyclohexane (Note 3) is placed in a 1.5-liter beaker and stirred magnetically. The immersion well is inserted into this solution and the lamp is started. The irradiation is continued until the conversion is judged to be essentially complete (Note 4). The reaction mixture is transferred to a 2-liter round-bottomed flask and is evaporated to dryness using a rotary evaporator and a water aspirator. The residue is dissolved in 50 ml of warm cyclohexane (Note 3), and the solution (Note 5) is poured onto a column of alumina (Note 6) 1.8 cm in diameter and about 8 cm in length. The round-bottomed flask is rinsed with three 10-ml portions of cyclohexane, and the rinsings are poured onto the column. The column is eluted with cyclohexane (100–150 ml) until further elution fails to yield additional 3-fluorophenanthrene in the eluate. The elution of any yellow material from the column should be avoided. The total eluate is evaporated to dryness, using a rotary evaporator, to give 1.92 g (98%) of white crystalline material. Recrystallization of this material from 8–10 ml of methanol gives 1.48 g (76%) of 3-fluorophenanthrene, m.p. 88.2–89.0° (lit.[2] m.p. 89.3–90.3°) (Note 7).

2. Notes

1. The use of various other types of mercury lamps and reaction vessels should also lead to satisfactory results, although low-pressure lamps emitting mainly at 254 nm are not recommended because inner filtering by the reaction product is a problem at this wavelength.

2. *trans-p*-Fluorostilbene can be prepared by a Grignard reaction involving *p*-fluorophenylmagnesium bromide and phenylacetaldehyde, followed by acid-catalyzed dehydration of the resulting carbinol.[2] Several alternative methods are available, including a Grignard reaction involving *p*-fluorobenzylmagnesium chloride and benzaldehyde with subsequent dehydration of the carbinol.

3. Redistilled Eastman Organic Chemicals practical grade cyclohexane can be used.

4. The irradiation time required depends on the type of light source used; with the 100-watt lamp used by the submitters, only negligible amounts of reactant were present after 10 hours of irradiation. The course

of the reaction can be followed conveniently by gas-liquid chromatography using any of a number of common column materials (e.g., a silicone gum rubber such as SE-52 or SE-30, or neopentyl glycol succinate). Three peaks are observed at intermediate stages in the conversion; in order of increasing retention time these correspond to *cis-p*-fluorostilbene, *trans-p*-fluorostilbene, and 3-fluorophenanthrene. If the conversion becomes inordinately slow during its final stages, it is advisable to interrupt the irradiation and remove any insoluble material that may have deposited on the immersion well.

5. The solution may be purple in color owing to incomplete removal of iodine during the reduced pressure evaporation.

6. Merck 71707 aluminum oxide can be used.

7. The checkers obtained material with m.p. 86.5–88° in 54% yield.

3. Methods of Preparation

This preparation is based on a procedure published by the submitters.[2] 3-Fluorophenanthrene has also been prepared previously from 3-phenanthrylamine *via* the diazonium fluoroborate.[3]

4. Merits of the Preparation

This general method has been used successfully in the submitters' laboratory in the preparation of about one hundred variously substituted phenanthrenes with such substituents as CH_3, $(CH_3)_3C$, C_6H_5, CN, CO_2H, CF_3, CH_3O, F, Cl, and Br. Other ring systems have also been prepared: chrysenes from α-styrylnaphthalenes; benzo[c]phenanthrenes from β-styrylnaphthalenes; picenes from di-α-naphthylethylenes; and benzo[c]chrysenes from 1-α-naphthyl-2-β-naphthylethylenes.

References

1. F. B. Mallory and C. S. Wood, *Org. Syn.*, **45**, 91 (1965).
2. C. S. Wood and F. B. Mallory, *J. Org. Chem.*, **29**, 3373 (1964).
3. P. M. G. Bavin and M. J. S. Dewar, *J. Chem. Soc.*, 4486 (1955).

3,4-HEXADIENOIC ACID

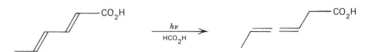

Submitted by C. F. MAYER and J. K. CRANDALL*
Checked by T. D. ROBERTS†

1. Procedure

A solution of 20 g (0.18 mole) of *trans,trans*-2,4-hexadienoic acid (Note 1) and 3.0 ml of formic acid (98–100%) in 1.2 liters of anhydrous ether in a 1.2-liter photolysis cell is degassed (Note 2) and then irradiated with a Vycor-filtered 450-watt Hanovia type L mercury arc in a quartz immersion well for approximately 25 hours when analysis by ultraviolet spectroscopy (Note 3) indicates that less than 5% of the initial amount of sorbic acid remains. The ether is removed at reduced pressure to give 23–24 g of viscous yellow oil. This oil is transferred to a 50-ml flask with ether washings for distillation. The ether and formic acid are removed at reduced pressure, and the residue is distilled to obtain 4.82–4.87 g of light yellow liquid, b.p. 84–100° (0.45 mm). Redistillation gives 4.09–4.31 g (20–22%) of 3,4-hexadienoic acid, b.p. 62–67° (0.3 mm) (Note 4).

2. Notes

1. The *trans,trans*-2,4-hexadienoic acid (sorbic acid) was obtained from Aldrich Chemical Company and used without purification. Gas chromatographic analysis (Note 5) of a sample esterified with diazomethane indicated that only a single component was present.

2. Degassing was performed by bubbling prepurified nitrogen through the solution in the photolysis cell for 1 minute using a glass tube with a fritted-glass tip. The immersion well was immediately inserted and the system was kept under a slight positive nitrogen pressure throughout the photolysis.

3. *Trans,trans*-2,4-hexadienoic acid has λ_{max} (EtOH) 263 nm, ε 25,800. The *cis,trans* isomer has λ_{max}(EtOH) 260 nm, ε 16,200.[1] Photochemical *cis-trans* isomerization of the sorbic acid apparently precedes formation of

* Indiana University, Bloomington, Indiana 47401.
† University of Arkansas, Fayetteville, Arkansas 72701.

the allene, since equilibration to the mixture of three isomers in about equal amounts is observed prior to the appearance of the allene (according to glpc assay of diazomethane-esterified photolysis aliquots). Quantitative evaluation of the data conservatively used ε 16,000. End absorption tailing into the region of interest further complicates the analysis, but this calculation gives a maximum concentration of sorbic acid. Alternatively, the photolysis can be followed by glpc analysis of diazomethane-esterified aliquots. The product and isomers of the starting material are the only significant peaks in the glpc trace. After 25 hours the 3,4-hexadienoate makes up about 90% of this mixture.

4. The acid was determined by glpc analysis of an esterified sample to be 96–98% pure. The infrared spectrum shows typical acid bands at 3–4 μ (OH) and 5.84 μ (C=O) and a weak allene peak at 5.06 μ. The nmr spectrum shows a sharp one-proton singlet at δ 11.92 (CO_2H), a two-proton multiplet at 5.2 (CH—C—CH), a two-proton multiplet at 3.1 (CH_2), and a three-proton multiplet at 1.7 (CH_3). Spin-decoupling with saturation of the δ 5.2 absorption reduced the multiplets at 3.1 and 1.7 to singlets.

5. Gas chromatography (glpc) was carried out on a Varian-Aerograph Series 1200 chromatograph using a 10 ft x ⅛ in. 15% Carbowax 20M on 60-80 Chromosorb W column.

3. Methods of Preparation

The procedure described here is essentially that of Crowley[2] and is the only method that has been used to prepare 3,4-hexadienoic acid.

4. Merits of the Preparation

This reaction is an interesting example of the deconjugation reaction whereby an α,β-unsaturated carbonyl compound is transferred into its β,γ-isomer.[3] The generally accepted intramolecular transfer of a γ-hydrogen atom involves transfer of a normally unreactive vinylic hydrogen atom in this case. Synthetically, this procedure provides a simple, one-step preparation of pure 3,4-hexadienoic acid from a readily available starting material. The presence of the reactive acid group in the allenic product makes this compound a likely source of new functionalized allenes for further study of the properties of this interesting functional group.

References

1. A. I. Scott, *Interpretation of the Ultraviolet Spectra of Natural Products*, The Macmillan Co., New York, New York, 1964, p. 81.

2. K. J. Crowley, *J. Amer. Chem. Soc.*, **85**, 1210 (1963).
3. N. C. Yang and M. J. Jorgenson, *Tetrahedron Lett.*, 1203 (1963); R. R. Rando and W. von E. Doering, *J. Org. Chem.*, **33**, 1671 (1968).

3-HYDROPEROXY-2,3-DIMETHYL-1-BUTENE

$$(CH_3)_2C=C(CH_3)_2 \xrightarrow[\text{sensitizer}]{h\nu/O_2} CH_2=C(CH_3)-C(CH_3)_2-OOH$$

Submitted by C. S. FOOTE and G. UHDE*
Checked by T. D. ROBERTS and L. MUNCHAUSEN†

1. Procedure (Note 1)

A solution of 8.4 g (0.1 mole) of 2,3-dimethyl-2-butene (Note 2), and 0.060 g of Rose Bengal (6.2×10^{-5} mole) (Note 3) in 150 ml of abs. methanol is irradiated with a 500-watt lamp (Sylvania, Iodine/Quartz 500T 3Q/Cl/U 120V) in a water-cooled (Note 4) immersion apparatus (Pyrex) through which oxygen is circulated by a pump (Dyna-vac, Cole-Palmer Instrument & Equipment Co.); O_2 uptake is followed by a gas buret. During 50–55 minutes, 2.32 liters of O_2 (0.095 mole, STP) is absorbed, after which time oxygen uptake ceases abruptly (Note 5). The solvent is removed with a rotary evaporator at 5–10° (Note 6). The liquid residue (11.3–11.8 g) is distilled (short path) to give 10.0–10.5 g (86–90%) of 3-hydroperoxy-2,3-dimethyl-1-butene, b.p. 54–56° (12 mm), n^{20}D 1.4420–1.4427 (Note 7).

2. Notes

1. The procedure is essentially that of Schenck and Schulte-Elte.[1]
2. 2,3-Dimethyl-2-butene may be prepared by dehydration of 2,3-dimethylbutan-2-ol[1,2] or from pinacol by way of the cyclic orthoformate.[3] The material used here was distilled (Vigreux column) and was about 97% pure, as shown by gas chromatography.
3. Technical grade Rose Bengal obtained from Eastman Kodak was used. A dye concentration of less than 10^{-4} M should be used, as higher

---

* Department of Chemistry, University of California, Los Angeles, California 90024. G. Uhde was on leave from Firmenich and Cie, Geneva, Switzerland.
† University of Arkansas, Fayetteville, Arkansas 72701.

concentrations lead to more rapid bleaching. Bleaching can be partially inhibited by adding a small amount of base (100 mg of Na_2CO_3 in 1–2 ml of water or a small amount of 0.01–0.001 M methanolic NaOH). Other dyes such as Eosin Y or Methylene Blue can be used, but offer no advantage and require slightly longer irradiation times.

4. The checkers noted that ordinary tungsten lamps of equal wattage are without merit for this purpose. Manufacturer's operating instructions for the lamp should be carefully followed.

5. The temperature of the reaction mixture is kept fairly constant (18–20°) with adequate cooling.

6. Use of a safety shield for this and subsequent operations is recommended. No difficulties have been encountered in numerous preparations of this compound, however.

7. The reported physical constants are: b.p. 54–55° (12 mm), $n^{20}D$ 1.4425.[1]

3. Methods of Preparation

3-Hydroperoxy-2,3-dimethyl-1-butene has also been prepared by addition of positive halogen sources to solutions of 2,3-dimethyl-2-butene and dilute hydrogen peroxide in alcohol (the "singlet oxygen" method).[4] In addition (and not to be confused with this method), positive halogen sources added to solutions of 2,3-dimethyl-2-butene with 98% H_2O_2 in ether produce a 2-halo-3-hydroperoxy-2,3-dimethylbutene which is smoothly dehydrohalogenated with base.[5]

4. Merits of the Preparation

The photosensitized oxygenation is a general route to allylic hydroperoxides. The double bond shifts cleanly to the allylic position; the reaction has been reviewed.[6] The method is superior to either of the previously mentioned techniques because even much less reactive olefins than the present one can be oxygenated in high yield on sufficiently long reaction, whereas the "singlet oxygen" method gives low yields of product with such acceptors,[7] although it works well in the present case.[4] The halohydroperoxide method is not general, since only tetrasubstituted olefins give allylic hydroperoxides, and less substituted olefins undergo cleavage.[5]

References

1. G. O. Schenck and K.-H. Schulte-Elte, *Ann.*, **618**, 185 (1958).
2. L. Henry, *Compt. Rend.*, **144**, 552 (1907).

3. G. Cranck and F. W. Eastwood, *Austr. J. Chem.*, **17**, 1392 (1964).
4. C. S. Foote, S. Wexler, W. Ando, and R. Higgins, *J. Amer. Chem. Soc.*, **90**, 975 (1968).
5. K. R. Kopecky, J. H. van de Sande, and C. Mumford, *Can. J. Chem.*, **46**, 25 (1968).
6. K. Gollnick, *Advan. Photochem.*, **6**, 1 (1968).
7. C. S. Foote, S. Wexler, and W. Ando, *Tetrahedron Lett.*, 4111 (1965).

ISOPHORONE DIMERS

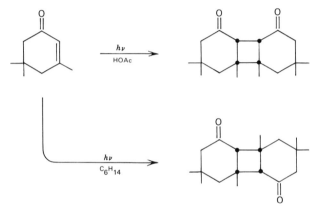

Submitted by D. J. TRECKER,* A. A. GRISWOLD,*
and O. L. CHAPMAN†
Checked by W. L. DILLING‡

1. Procedure

A. Head-to-Head Isomer. The apparatus consists of a tubular Pyrex reactor, 15 in. long and 6.9 cm. O.D., which contains an internal, concentric, water-jacketed Pyrex immersion well, 10 in. long and 4.6 cm. O.D. The reactor is equipped with a reflux condenser, a thermocouple well, a septum-capped side arm for sample withdrawal, and a fritted-glass diffuser at the bottom of the reactor for nitrogen ebullition. The light source is a 450-watt medium-pressure Hanovia mercury arc which is suspended in the immersion well.

* Research and Development Department, Union Carbide Corporation, Chemicals and Plastics, South Charleston, West Virginia 25303.
† Iowa State University, Ames, Iowa.
‡ The Dow Chemical Company, Midland, Michigan 48640.

A solution of 70 g of isophorone (0.51 mole) (Note 1) in 90 volume percent aqueous acetic acid (500 ml) is irradiated under a gentle nitrogen sparge (Note 2) for 20 hours. During this time the water flow on the lamp jacket is maintained at such a rate that the reaction temperature is held at 30–40°.

At the end of the irradiation period the reaction solution is placed in a large beaker, and water is added until the cloud point is reached. Upon standing for several hours (conveniently overnight), a mixture of dimers crystallizes from solution and is collected by filtration. The crystals are washed with three 100-ml portions of n-heptane and are dried in a vacuum oven (50°) overnight. The total yield is 22 g (0.08 mole), or 31.4%, based on charged isophorone. The composition is analyzed (Note 3) to be 83% head-to-head dimer[1] (Note 4) and a 17% mixture of head-to-tail dimers[1] (Note 5).

Pure head-to-head dimer is obtained by stirring for 17 hours at room temperature the mixture of isomers (22 g, 0.08 mole) with concentrated sulfuric acid (60 g) (Note 6). The acidolysis solution is then poured into 300 g of cracked ice. After the ice entirely melts, the precipitated solid is collected by filtration, washed successively with two 300-ml portions of water and two 200-ml portions of hot n-heptane, and dried in a vacuum oven (50°). The resulting white powder (13 g) melts at 188–189°[1] and consists of 98.5–99.2% pure head-to-head dimer (Notes 3, 7).

B. Head-to-Tail Isomer. In the same apparatus described above, a solution of 70 g of isophorone (0.51 mole) in 500 ml of n-hexane is irradiated for 88 hours (Note 8). At the end of that period the precipitated crystals are collected by filtration, washed with two 50-ml portions of n-hexane, and dried. Further purification may be achieved by sublimation (125°/0.4 mm). Total yield is 3.8 g (0.014 mole), or 5.4%, based on charged isophorone. The white needles melt at 211–212° (evacuated capillary) and consist of 98.5–99.0% pure head-to-tail dimer[1] (Note 9).

2. Notes

1. Commercial grade isophorone may be used without further purification. The material is obtained from Union Carbide Corp., Chemicals and Plastics, South Charleston, West Virginia.

2. Linde H.P. dry nitrogen may be used directly without further purification. Nitrogen of lower purity should be passed through a Fieser train[2] before use.

3. The isomer distribution is analyzed by gas chromatography, employing a 10 ft × 0.25 in. column of 5% Carbowax 20M on acid-washed,

dichlorodimethylsilane-treated Chromosorb G, at a gas flow of 120 ml of helium per minute, and at a temperature of 225°.

4. cis,syn,cis-5,5,7,8,10,10-Hexamethyltricyclo[6.4.0.0^{2,7}]dodecane-3,12-dione.

5. cis,syn,cis- and cis,anti,cis-1,5,5,7,11,11-Hexamethyltricyclo[6.4.0.0^{2,7}]dodecane-3,9-dione.

6. This acidolysis selectively dedimerizes the head-to-tail isomers.

7. Recrystallization from hot *n*-heptane (*ca.* 900 ml) affords white needles but does not improve the purity as determined by either melting point or vpc analysis.

8. Continued irradiation provides no additional dimer formation.

9. The exact stereochemical assignment (*cis,syn,cis* or *cis,anti,cis*) (Note 5) has not been made at this time. The second head-to-tail dimer is considerably more soluble in nonpolar media, and consequently its isolation requires an elaborate chromatographic procedure.

3. Methods of Preparation

Photodimerization of cyclic olefins, particularly conjugated enones, is the only convenient method for preparing a broad range of 6,4,6-fused ring compounds.[3] The isophorone dimers cannot be readily prepared by other means.

4. Merits of the Preparation

The unique use of solvent to direct the dimerization selectively to either head-to-head or head-to-tail isomers makes this an extremely simple procedure, particularly with regard to workup and isolation of the products. Moreover, isophorone is the most readily available and least expensive of the cycloalkenones and, hence, provides a convenient entry into the 6,4,6-fused ring system.

References

1. O. L. Chapman, P. J. Nelson, R. W. King, D. J. Trecker, and A. A. Griswold, *Rec. Chem. Progr.,* **28,** 167 (1967).
2. L. F. Fieser, *Experiments in Organic Chemistry,* 2nd ed., D. C. Heath and Co., Boston, 1941, pp. 395–396.
3. W. L. Dilling, *Chem. Rev.,* **66,** 373 (1966); R. Steinmetz, *Fortsch. chem. Forsch.,* **7,** 445 (1967).

6-*epi*-LUMISANTONIN

Submitted by D. I. SCHUSTER and A. C. FABIAN*
Checked by F. B. MALLORY and C. W. MALLORY†

1. Procedure

Gaseous hydrogen chloride is added to 800 ml of dimethyl formamide in a 2-liter round-bottomed flask to make a 5% solution. To this is added 80 g of α-santonin (Note 1) and the solution is heated at 100° for 8 hours. The orange reaction mixture is added to ice, and 600 ml of water is added. The solution is repeatedly extracted with ethyl ether until the ether layer is colorless, requiring about 2 liters of ether. The ether is washed with two 100-ml portions of 10% potassium hydroxide solution, affording a blood-red aqueous layer (Note 2), washed several times with water, and then dried over magnesium sulfate. Removal of solvent with a water aspirator affords 45 g of a reddish white solid which is recrystallized several times from ethyl acetate to give 40 g (50%) of white crystals, m.p. 104–105° (Note 3) of 6-*epi*-α-santonin.

A solution of 4.3 g of 6-*epi*-α-santonin in 200 ml of Spectrograde dioxane (Note 4) is irradiated at 2537 Å in a Rayonet Srinivasan-Griffin Chamber Reactor using RPR-2537 lamps and a quartz reaction vessel RQV 218 (Notes 5, 6). The progress of reaction is monitored by disappearance of the infrared band at 1660 cm^{-1} due to starting materials, and appearance of a new band at 1705 cm^{-1} due to product (Note 7). The reaction is complete after 9 hours of irradiation (Note 8). The solvent is

---

* New York University, University Heights, New York, New York, 10453.
† Bryn Mawr College, Bryn Mawr, Pennsylvania 19010.

removed using an aspirator, and the residual yellow oil, 4.9 g, is chromatographed on 120 g of silica gel (Note 9) on a column prepared with 3:1 benzene-cyclohexane. The column is eluted with 2 liters of benzene and 1.5 liters of 2% ether-benzene to give a yellow oil containing negligible amounts of product. Further elution with 5 liters of 5% ether-benzene gives a white solid which is recrystallized from absolute ethanol to give 2.1 g (49%) of 6-*epi*-lumisantonin, m.p. 135–138° (Note 10). The ultraviolet spectrum in 95% ethanol has λ_{xam} 241 nm, ε 4160.

2. Notes

1. α-Santonin obtained from Mann Research Laboratories was used.
2. Additional washing with aqueous base lowers the yield of product.
3. Reported m.p. 104°.[1,2]
4. The checkers used Fisher Certified dioxane and obtained the product without change in yield.
5. Reactor, lamps, and reaction vessel were obtained from Southern New England Ultraviolet Company.
6. The checkers did not have the apparatus described, and they improvised an equally effective low-pressure light source and reaction system. They used a single 8-watt General Electric germicidal lamp (*ca.* 12 in. long and ⅝ in. in diameter) mounted vertically alongside a reaction vessel consisting of a simple quartz tube 18 in. long and 20 mm. in diameter and sealed at one end. The lamp and reaction tube were loosely wrapped with aluminum foil. With this arrangement, 1 g of 6-*epi*-α-santonin in 70 ml of dioxane required 9 hours of irradiation. At least four such reaction tubes could be placed around the lamp, permitting parallel irradiation on the scale given by the authors.
7. Infrared spectra in chloroform solution.
8. Care must be taken to avoid overirradiation, which leads to conversion of the product into additional photoproducts.[3]
9. The authors used Davison grade 950 silica gel, 60–200 mesh. The checkers used Davison grade 12 silica gel. In the latter case, it was necessary to elute with 1 liter of ether to remove all of the desired product from the column.
10. The checkers obtained m.p. 134.8–135.8°.

References

1. H. Ishikawa, *Bull. Phar. Soc. Japan,* **76,** 504 (1956).
2. D. H. R. Barton, J. E. D. Levisalles, and J. T. Pinhey, *J. Chem. Soc.,* 3472 (1962).
3. D. I. Schuster and A. C. Fabian, *Tetrahedron Lett.,* 4093 (1966).

3-METHOXY-4-AZATRICYCLO[3.3.2.0^{2,8}]DECA-3,6,9-TRIENE

Methoxyazabullvalene

Submitted by LEO A. PAQUETTE, GRANT R. KROW, and THOMAS J. BARTON*
Checked by JOHN R. MALPASS†

1. Procedure

A. 7-Azabicyclo[4.2.2]deca-2,4,9-trien-8-one. Into a 1-liter three-necked flask equipped with a magnetic stirrer, a 125-ml pressure-equalizing dropping funnel, and a reflux condenser capped with a calcium chloride drying tube is placed 52 g (0.5 mole) of cyclooctatetraene (Note 1). With stirring, the hydrocarbon is heated to 50° by means of an external electrically heated oil bath, and chlorosulfonyl isocyanate (56.5 g, 0.4 mole) (Note 2) is added dropwise during 1 hour. The reaction mixture darkens after approximately 20 minutes, at which time the current to the oil bath is stopped as necessary in order to maintain a temperature of 50–55° (Note 3). After 4 hours at this temperature, the reaction mixture is permitted to cool during several hours and a solid cake forms (Note 4). The dark solid *N*-(chlorosulfonyl)lactam is dissolved in 100–150 ml of acetone, and this solution is added dropwise to a mixture of 200 ml of water and 100 ml of acetone while maintaining pH 7 by concurrent dropwise addition of 4 *N* sodium hydroxide solution (Note 5).

The precipitated sodium sulfate is filtered and washed with small portions of methylene chloride. The filtrate is extracted with three 100-ml

---

* Ohio State University, Columbus, Ohio 43210.
† University of Leicester, Leicester, England.

portions of methylene chloride, and the combined washings and extracts are dried over anhydrous magnesium sulfate, filtered, and evaporated under reduced pressure. The crude 7-azabicyclo[4.2.2]deca-2,4,9-trien-8-one (39–52 g, 66–88%) is collected as a white solid, m.p. 137–140°. Recrystallization of this product from acetone-ether gives white crystals, m.p. 139–140°.

B. *7-Methoxy-8-azabicyclo[4.2.2]deca-2,4,7-9-tetraene.* The lactam from the previous step (15 g, 0.10 mole) is added in one portion to an ice-cooled solution of trimethyloxonium fluoroborate (16.0 g, 0.11 mole) (Note 6) in 200 ml of dry methylene chloride contained in a 500-ml., one-necked flask previously fitted with a magnetic stirrer and a calcium chloride drying tube. The stirred solution is allowed to warm to room temperature during 1 hour and stirring is continued for 5–7 hours.

A 50% aqueous solution of potassium carbonate (16 ml) is then added carefully and after 1 hour the salts are removed by filtration. The filtrate is dried over anhydrous magnesium sulfate, filtered, and evaporated to afford a reddish oil. This oil is extracted with 250 ml of warm petroleum ether (60–80°), and the solution is decolorized by filtration through a small column of neutral alumina. Concentration of the eluate to small volume and cooling to 0° yields 11–13.5 g (68–84%) of the imino ether, m.p. 50.5–52°. The product may also be purified by sublimation at 55° (0.3 mm).

C. *3-Methoxy-4-azatricyclo[3.3.2.0^{2,8}]deca-3,6,9-triene.* The imino ether (2.4 g, 0.015 mole) is dissolved in 500 ml of reagent grade acetone (Note 7), and this solution is irradiated in a quartz immersion well fitted with an immersion-type 450-watt Hanovia mercury arc (Type L) and a Pyrex filter sleeve. The photolysis, which can be conveniently monitored by gas chromatographic analysis of small aliquots (Note 8), generally requires 6–8 hours for completion. Removal of the solvent under reduced pressure affords a yellow oil which is dissolved in 50–75 ml of petroleum ether (60–80°) and clarified by filtration through neutral alumina. Concentration affords a colorless oil which is crystallized from petroleum ether to give 1.7–2.1 g (71–88%) of white crystals, m.p. 49.5–50.5°. Methoxyazabullvalene may be purified further by sublimation.

2. Notes

1. The cyclooctatetraene employed in this work was generously supplied by Badische Anilin und Soda Fabrik. The hydrocarbon need not be redistilled prior to reaction.

2. The submitters used chlorosulfonyl isocyanate as supplied by Ameri-

can Hoechst Corp., New York, New York. Alternatively, this reagent may be prepared by a procedure which is described elsewhere.[1]

3. The reaction is exothermic and the oil bath will usually remain at 50° for 2–3 hours because of the heat of the reaction. *Caution!* On two occasions, overheating occurred causing resinification of the reaction mixture and spewing of the tarry material about the laboratory. Therefore the operator should perform this experiment behind an appropriate safety shield in an efficient hood.

4. If the mixture fails to solidify, it has proven advantageous to reheat the liquid to 50° for 2–3 additional hours.

5. The aqueous alkali may be added from a buret and the change in pH conveniently followed by use of a pH meter.

6. Optimum yields are obtained only with freshly prepared trimethyloxonium fluoroborate.[2]

7. Under these conditions, acetone absorbs approximately 80% of the incident light at 280 mμ.

8. Gas chromatographic analyses were performed with the aid of an Aerograph Hy-Fi Model 600-D gas chromatographic unit fitted with a 10 ft $\times$ 0.125 in. column packed with 5% SE-30 on Chromosorb W.

3. Methods of Preparation

This preparation of methoxyazabullvalene is based upon the published procedure of the submitters,[3,4] and is at present the only known synthetic route to the azabullvalene system.

4. Merits of the Preparation

The photochemical step of the sequence has the merit of being quite general for the synthesis of bullvalene[5,6] and a variety of azabullvalenes. This four-step procedure provides an exceptionally facile synthetic entry to a theoretically interesting group of heterocycles.

References

1. R. Graf, *Org. Syn.*, **46**, 23 (1966).
2. H. Meerwein, *Org. Syn.*, **46**, 120 (1966).
3. L. A. Paquette and T. J. Barton, *J. Amer. Chem. Soc.*, **89**, 5480 (1967).
4. L. A. Paquette, T. J. Barton, and E. B. Whipple, *J. Amer. Chem. Soc.*, **89**, 5481 (1967).
5. M. Jones, Jr., and L. T. Scott, *J. Amer. Chem. Soc.*, **89**, 150 (1967).
6. W. von E. Doering and J. W. Rosenthal, *Tetrahedron Lett.*, 349 (1967).

2-METHOXY-5-HYDROPEROXY-2,5-DIMETHYLDIHYDROFURAN

$$\text{furan} \xrightarrow[\text{CH}_3\text{OH}]{h\nu/O_2,\ \text{sensitizer}} \text{2-methoxy-5-hydroperoxy-2,5-dimethyldihydrofuran (OOH, OCH}_3\text{)}$$

Submitted by C. S. FOOTE and G. UHDE*
Checked by T. D. ROBERTS and L. MUNCHAUSEN†

1. Procedure (Note 1)

A solution of 9.6 g (0.10 mole) of 2,5-dimethylfuran (Note 2) and 0.060 g of Rose Bengal (6.2 x 10^{-5} mole) (Note 3) in 150 ml of absolute methanol is irradiated with a 500-watt tungsten lamp (Sylvania, iodine-quartz 500 T 3Q/Cl/U) (Note 4) in a water-cooled Pyrex immersion apparatus (Note 5) through which oxygen is circulated by a diaphragm pump (Dyna-vac, Cole-Palmer Instrument & Equipment Co.). Oxygen uptake is monitored by a gas buret. In 45–50 minutes the oxygen uptake abruptly stops after 2.34 liters of oxygen (0.096 mole STP) has been absorbed.

The solvent is removed with a rotary evaporator at 5–10° (15 mm) (Note 6). The residue is dissolved in about 100 ml of ether, washed with iced 2% aqueous bicarbonate, followed by ice water (Note 7) until most of the dye is removed, and dried over magnesium sulfate.

The ether is distilled under reduced pressure at 5–10°, and the crude crystalline material (14.8–15.2 g) yields 11.8–13.1 g (74–82%) of colorless crystals of 2-methoxy-5-hydroperoxy-2,5-dimethyldihydrofuran, m.p. 74–75° (Note 8), after recrystallization from ether.

2. Notes

1. The method is that of Foote et al.[1]
2. 2,5-Dimethylfuran was obtained from K & K Laboratories, Inc. The material as received contains a large mount of ketonic impurity. After filtration twice through silica gel and distillation, the material is 96% pure by gas chromatography.

---

* Department of Chemistry, University of California, Los Angeles, California 90024. G. Uhde was on leave from Firmenich and Cie, Geneva, Switzerland.
† University of Arkansas, Fayetteville, Arkansas 72701.

3. Technical grade Rose Bengal obtained from Eastman Kodak was used. A concentration of less than 10^{-4} M dye should be used, as higher concentrations lead to more rapid bleaching. Bleaching can be partially inhibited by adding a small amount of base (100 mg of Na_2CO_3 in 1–2 ml of water or a small amount of 0.01–0.001 M methanolic NaOH). Other dyes such as Eosin Y or Methylene Blue can be used, but offer no advantage, and require slightly longer irradiation times.

4. The checkers noted that ordinary tungsten lamps of equal wattage are without merit for this purpose. Manufacturer's operating instructions for the lamp should be carefully followed.

5. The temperature of the reaction mixture is kept fairly constant (18–20°) with adequate cooling.

6. *Caution!* All further manipulations (concentration of the peroxides, purification, etc.) are carried out behind an effective shield, and safety gloves are worn. However, the product does not seem to be particularly sensitive and can even be sublimed under reduced pressure.[1]

7. As the methoxy hydroperoxide is easily hydrolyzed, the wash should be as rapid as possible, using small portions of cold water and bicarbonate solution.

8. The melting point reported[1] was 75–76.°

3. Methods of Preparation

2-Methoxy-5-hydroperoxy-2,5-dimethyldihydrofuran has also been prepared by addition of aqueous hypohalite to solutions of dimethylfuran and H_2O_2 in methanol.[2]

4. Merits of the Preparation

The photooxygenation of dimethylfuran illustrates a generally applicable method of oxidizing furans, dienes, linear polycyclic aromatic compounds of the anthracene series and higher, and many other types of heterocyclic compounds.[3] The products in each case appear to be derived from an *endo*-peroxide formed by addition of oxygen in a 1,4 fashion across the diene system; in the case of heterocyclic compounds, the initial peroxide undergoes subsequent reactions (in this case, addition of solvent). The method is more generally applicable than the hypochlorite-H_2O_2 method (which leads to the same products) because even unreactive compounds can be oxidized in high yield upon sufficiently long irradiation; the hypochlorite-H_2O_2 method[2] is limited by the large excesses of reagents which must be used with inefficient acceptors.

References

1. C. S. Foote, M. T. Wuesthoff, S. Wexler, I. G. Burstain, R. Denny, G. O. Schenck, and K.-H. Schulte-Elte, *Tetrahedron*, **23**, 2583 (1967).
2. C. S. Foote, S. Wexler, W. Ando, and R. Higgins, *J. Amer. Chem. Soc.*, **90**, 975 (1968).
3. K. Gollnick and G. O. Schenck, in J. Hamer, Ed., *1,4 Cycloaddition Reactions*, Academic Press, New York, 1967, p. 255.

trans-9-METHYL-1,2,3,4,4a,9a-HEXAHYDROCARBAZOLE

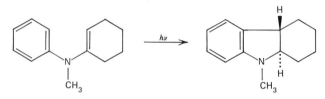

Submitted by G. L. EIAN and O. L. CHAPMAN*
Checked by T. D. ROBERTS†

1. Procedure

A. 1-(N-Methylanilino)cyclohexene. Cyclohexanone diethyl ketal[1] (20 g, 0.116 mole) and freshly distilled N-methylaniline (12.5 g, 0.116 mole) are placed in a small round-bottomed flask equipped with a magnetic stirring bar and fitted with a vacuum, fractional-distillation head. A trace of p-toluenesulfonic acid is added, and the mixture is heated in a wax bath until ethanol begins to distill over (bath temperature about 150°). The bath temperature is slowly raised over a period of 3 hours to 200°. The system is then protected with a calcium chloride drying tube and cooled. Fractional distillation of the residue under vacuum gives 1-(N-methylanilino)cyclohexene (16 g, 75%), b.p. 105–108° (1.5 mm), lit.[2] 140° (13 mm).

B. trans-9-Methyl-1,2,3,4,4a,9a-hexahydrocarbazole (Note 1). A solution of 1-(N-methylanilino)cyclohexene (4.20 g, 2.36 x 10⁻² mole) in ether (300 ml) from a freshly opened can is purged with argon for 30 minutes and irradiated for 6 hours with a Hanovia type A 550-watt mercury arc

* Iowa State University, Ames, Iowa 50010.
† University of Arkansas, Fayetteville, Arkansas 72701.

lamp in a Pyrex immersion well. Care must be taken to keep the system anhydrous because of the sensitivity of the enamine. The reaction is monitored by thin layer chromatography using Silica Gel H and eluting with ethyl acetate in petroleum ether (Skelly B). Evaporation of the solvent after irradiation leaves an oil which contains *trans*-9-methyl-1,2,3,4,4a,9a-hexahydrocarbazole (71%), *cis*-9-methyl-1,2,3,4,4a,9a-hexahydrocarbazole (3%), and a trace of 9-methyl-1,2,3,4,-tetrahydrocarbazole (Note 2). Crystallization of the crude product from 95% ethanol gives *trans*-9-methyl-1,2,3,4,4a,9a-hexahydrocarbazole (2.3 g, 55%), m.p. 58–60° (lit.[3] 59–61°), picrate m.p. 125–126° (lit.[3] 122–124°), methiodide m.p. 233–234° (lit.[3] 233–234°).

2. Notes

1. For related examples of this cyclization see Chapman and co-workers.[4]
2. Analysis was performed by vpc using a 5 ft by 0.25 in. column of 5% potassium hydroxide and 10% polyethylene glycol (Carbowax 20M) on 60/80 mesh Chromosorb W. Benzophenone was added as an internal standard, and corrections were made for differences in thermal conductivity.

References

1. U. Schmidt and P. Grafen, *Ann.*, **656**, 97 (1962).
2. J. Hoch, *Compt. Rend.*, **200**, 938 (1935).
3. T. Masamune, *Bull. Chem. Soc. Jap.*, **30**, 491 (1957).
4. O. L. Chapman and G. L. Eian, *J. Amer. Chem. Soc.*, **90**, 5329 (1968); O. L. Chapman, G. L. Eian, A. Bloom, and J. C. Clardy, *ibid.*, **93**, (1971).

N-METHYL-4-NITRO-o-ANISIDINE

$$\underset{(I)}{\text{OCH}_3, \text{OCH}_3, \text{NO}_2} \xrightarrow[\text{(DMSO/H}_2\text{O)}]{h\nu, \text{CH}_3\text{NH}_2} \underset{(II)}{\text{OCH}_3, \text{NHCH}_3, \text{NO}_2} + \underset{(III)}{\text{NHCH}_3, \text{OCH}_3, \text{NO}_2}$$

Submitted by C. FRÁTER and J. CORNELISSE*
Checked by J. A. VINK*

1. Procedure

4-Nitroveratrole (I) (1.36 g, 0.0074 mole) is dissolved in 140 ml of dimethylsulfoxide. To this solution 90 ml of a 40% methylamine solution in water (Notes 1, 2) is added. The mixture is irradiated with Pyrex-filtered light from a high-pressure mercury lamp (Hanau Q 81), while the temperature is kept at 18–20°. During the irradiation the solution is stirred continuously.

The irradiation is stopped when the absorbance of the reaction mixture at 262 nm is about 1.75–1.80 times as high as the absorbance at 334 nm at the beginning of the reaction (Note 3). In order to follow the absorbance of the reaction mixture during the irradiation, aliquots are taken at frequent intervals, diluted with ethanol, and measured. The time necessary to obtain this amount of conversion is about 3–4 hours (Note 4). The initially yellow solution turns dark red.

The solution is diluted with twice its volume of water and extracted six times with light petroleum (boiling range 60–80°). The combined light petroleum fractions are dried with anhydrous sodium sulfate. After filtration, the drying agent is washed with chloroform, and the chloroform added to the solution. The solvent is then evaporated at reduced pressure (Note 5). The remaining, more or less crystalline, substance is dissolved in as little chloroform as possible and subjected to column chromatography on 60 g of silica (Merck 0.05–0.2; column 25 mm I.D.) with 12% dichloromethane in hexane as elution liquid.

The N-methyl-4-nitro-o-anisidine (II), which is yellow-orange as a solid, appears on the column as a red band, running faster than the other

* Laboratory of Organic Chemistry, The University, Leiden, The Netherlands.

products. When this red band has reached the lower end of the column, further elution is performed with 25% dichloromethane in hexane. The chromatography takes about 10 hours.

After combining the chromatography fractions containing the desired product and evaporating the solvent, N-methyl-4-nitro-o-anisidine is obtained as an almost pure crystalline solid. Yield 900 mg (66%), m.p. 85.5–87°. Further purification can be carried out by recrystallization from light petroleum (60–80°).

2. Notes

1. It is important to prepare the solution in the indicated order; the 4-nitroveratrole does not readily dissolve in the methylamine solution, even when dimethylsulfoxide is added.

2. DMSO was obtained from J. T. Baker Chemical Co., and methylamine from Fluka, A.G. Chemische Fabrik.

3. This corresponds to 95% conversion of 4-nitroveratrole. Ultraviolet absorption data (in ethanol):

4-Nitroveratrole (I)	236 nm ($\varepsilon = $ 9010)
	301 nm (shoulder)
	334 nm ($\varepsilon = $ 6820)
N-Methyl-4-nitro-o-anisidine (II)	235 nm ($\varepsilon = $ 10550)
	262 nm ($\varepsilon = $ 16700)
	307 nm ($\varepsilon = $ 4500)
	395 nm ($\varepsilon = $ 3320)
N-Methyl-5-nitro-o-anisidine (III)	229 nm ($\varepsilon = $ 5100)
	263 nm ($\varepsilon = $ 4560)
	399 nm ($\varepsilon = $ 13180)

4. The reaction proceeds faster when the irradiation is performed in a Rayonet photochemical reactor using the 350-nm lamps.

5. At this stage the composition of the reaction mixture can be determined by vpc (1.5 m, SE-30, 180°). The retention times are in the order (I) < (II) < (III).

6. A product (IV) could be identified as the compound in which both OCH_3 groups were replaced by $NHCH_3$ groups. The yield is about 1%, m.p. 172°.

3. Discussion

This synthesis exemplifies the use of the characteristic meta activation that is often encountered in aromatic photosubstitutions.[1,2]

The reaction can also be performed with a 2.5 x 10^{-3} M solution of 4-nitroveratrole in 30% methanol-water containing 16% methylamine. Extraction with light petroleum is not necessary in this case. The solvent can be evaporated after the irradiation has been completed, and the remaining solid purified by column chromatography as described. With the Rayonet photochemical reactor, 96% conversion was obtained within 1 hour. The yield of N-methyl-4-nitro-o-anisidine was 73%.

References

1. M. E. Kronenberg, A. van der Heyden, and E. Havinga, *Rec. Trav. Chim.*, **86**, 254 (1967).
2. E. Havinga, R. O. de Jongh, and M. E. Kronenberg, *Helv. Chim. Acta*, **50**, 2550 (1967).

3-METHYL-2-OXATETRACYCLO[4.2.1.0^{3,9}.0^{4,8}]NONANE

Submitted by R. R. SAUERS, W. SCHINSKI, and B. SICKLES*
Checked by T. D. ROBERTS†

1. Procedure

A solution of 40.0 g (0.29 mole) of freshly distilled 5-acetylnorbornene (Note 1) and 20 g of freshly distilled piperylene in 1.1 liters of dry benzene is placed in a flask equipped with a gas inlet tube, reflux condenser, and an immersion well (Note 2). The solution is deoxygenated by passing a stream of nitrogen through the solution for 15 minutes. The nitrogen flow is decreased to a slow stream and irradiation is commenced (Note 3). After 48 hours the reaction is stopped; at this point most of the starting material has reacted (Note 4). The benzene and piperylene are removed by evaporation at reduced pressure and the residue is distilled at 97–100° (63 mm) to give 28 g (83% yield) of a colorless oil. The material obtained in this way is contaminated with 6.5% of the starting materials. If oxetane of higher purity is desired, several washings of an ethereal solution of

---

* Rutgers, The State University, New Brunswick, New Jersey.
† University of Arkansas, Fayetteville, Arkansas 72701.

the product with dilute potassium permanganate solution remove the starting materials without significant loss of oxetane. Purified oxetane has b.p. 60° (10 mm).

2. Notes

1. 5-Acetylnorbornene was prepared by the method of Dinwiddie and McManus[1] except that the reactants were cooled in dry ice before mixing. An *exo-endo* mixture (21–79) was obtained which was used without further separation.
2. A commercial water-cooled quartz well was used with a Corex (9700) filter sleeve.
3. The lamp used was a 450-watt Hanovia type L medium-pressure mercury source.
4. The reaction can be conveniently monitored by gas chromatography on a 12 ft × ¼ in. column of Carbowax 20M (10%) on Chromosorb G at 160°.

Reference

1. J. G. Dinwiddie, Jr., and S. P. McManus, *J. Org. Chem.*, **30**, 766 (1964).

2-METHYLENEBICYCLO[2.1.1]HEXANE

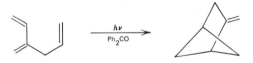

Submitted by L. SKATTEBØL and J. PEYMAN*
Checked by R. SRINIVASAN†

1. Procedure

A. 3-Methylene-1,5-hexadiene. A mixture of 1,2,6-heptatriene and tricyclo[4.1.0.0¹,³]heptane (18.8 g, 0.2 mole), obtained in 85% yield from 1,1-dibromo-2(3-butenyl)cyclopropane and methyllithium,[1] is placed in a 50-ml round-bottomed flask. This is connected through an adapter to a

---

* Union Carbide Research Institute, P.O. Box 278, Tarrytown, New York 10591.
† IBM Research Center, Yorktown Heights, New York 10598.

40-cm horizontally positioned tube, packed with Pyrex wool, and kept at 350° by an electrically heated oven (Note 1). At the other end of the tube a second 50-ml round-bottomed flask is attached through an adapter with a side arm, which is connected to a vacuum pump. This flask is cooled to —78° by means of a Dry Ice-acetone bath and the system is evacuated to 5 mm. The hydrocarbon mixture starts to boil, and after a short while a liquid commences to collect in the receiver. In order to maintain a reasonable rate, the distilling flask is surrounded by a water bath kept at about 50°. After approximately 1 hour the reaction is completed. Nitrogen is let into the system (Note 2) and the product allowed to attain room temperature. Gas chromatography shows the presence of >99% of 3-methylene-1,5-hexadiene (Note 3).[1,2] This product is used in the irradiation step without further purification. The yield is nearly quantitative.

B. *2-Methylenebicyclo[2.1.1]hexane.* A 100-ml irradiation vessel (Note 4) is charged with 9.4 g (0.1 mole) of 3-methylene-1,5-hexadiene, 940 ml of sodium-dried pentane, and 4.7 g of benzophenone. The solution is irradiated through a Pyrex immersion well which is cooled with tap water. The light source is a 450-watt high-pressure mercury vapor lamp (Note 5). A slow stream of nitrogen is passed through the reaction vessel during the irradiation.

After 20 hours, gas chromatography of an aliquot shows that the reaction is completed and that only one volatile product is present. The solution is filtered (Note 6) and the pentane carefully removed through a column (Note 7). The residue is distilled through a 10-cm Vigreux column giving 6.8–7.3 g (73–78% yield) of 2-methylenebicyclo[2.1.1]hexane, b.p. 94°, n^{24}D 1.4611 (Note 8).

2. Notes

1. The submitters used a 750-watt furnace, commercially available from Hevi-Duty Electric Co., Watertown, Wisconsin.

2. Air can be introduced but, in order to avoid condensing water in the product, it is advisable to pass it through a drying tube.

3. B.p. 56° (155 mm.), n^{23}D 1.4546. It should be stored under nitrogen in the cold.

4. The irradiation apparatus is commercially available from Ace Glass, Inc., Vineland, New Jersey (reaction vessel, cat. No. 6515-27; immersion well, cat. No. 6515-25). A detailed description of the apparatus is given in reference 3.

5. The irradiation lamp and power supply are available from Ace Glass,

Inc., (cat. Nos. 6515-34, 6515-60) and from Engelhard Hanovia, Inc., Newark, New Jersey (cat. Nos. 679A-36, 34245).
6. Some benzopinacol and polymeric material had formed.
7. The submitters used a 60-cm spinning-band column. Any efficient column could be used.
8. The preparation of 2-methylenebicyclo[2.1.1]hexane by this procedure has been reported.⁴

References

1. L. Skattebøl, *J. Org. Chem.*, **31**, 2789 (1966).
2. L. Skattebøl and S. Solomon, *J. Amer. Chem. Soc.*, **87**, 4506 (1965).
3. D. R. Arnold, *Advan. Photochem.*, **6**, 341–343 (1968).
4. P. de Mayo, J. L. Charlton, and L. Skattebøl, *Tetrahedron Lett.*, 4679 (1965).

trans-β-NITROSTYRENE DIMER*

Cyclobutane, trans-1,3-dinitro-trans-2,4-diphenyl-

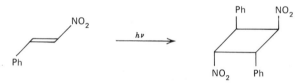

Submitted by DONALD G. FARNUM and ABDOL J. MOSTASHARI†
Checked by L. WILLIAMS and R. SRINIVASAN‡

1. Procedure

The photodimerization of β-nitrostyrene (Notes 1, 2) is conducted in an 18-in. open crystallizing dish cooled in a water bath to 18°. The starting material (76.5 g) is slurried in 800 ml of water and stirred vigorously with a mechanical stirrer fitted with a 2-in. Teflon paddle. A Hanovia-type Pyrex insert is placed almost horizontally 1 in. above the dish in such a manner that the length of the lamp inside the insert is over the suspension. A sheet of aluminum foil is used to cover the insert as a reflector. Irradia-

---

† Michigan State University, East Lansing, Michigan 48823.
‡ IBM Research Center, Yorktown Heights, New York 10598.
* This work was supported by the National Science Foundation under Grant GP10734.

tion is conducted for 24 hours with a 550-watt Hanovia lamp. The suspension is then filtered, air-dried, and washed by Soxhlet extraction with hexane to give a crude solid (59 g, 77%), m.p. 175–181°. Recrystallization of the crude material from benzene affords pure dimer, m.p. 183–185° (41.3 g, 54%) (lit.[1] 181–182°) (Notes 3, 4).

2. Notes

1. β-Nitrostyrene was prepared according to the procedure reported in *Organic Syntheses*[2] or was obtained by recrystallization of commercial material (Aldrich) from ethanol.
2. *Caution!* β-Nitrostyrene is a skin irritant; contact should be avoided. Recrystallization under a hood is advised.
3. *trans*-Cinnamic acid can also be photodimerized in this manner, but yields are generally lower.[3]
4. Yields of pure dimer varied from 45 to 65% for irradiation times of 36 hours.

3. Merits of the Preparation

The procedure of Campbell and Ostead[1] is suitable for 25–40% yields of small amounts of material. This modified method is applicable to the preparation of large quantities of material in good yield.

References

1. R. D. Campbell and R. F. Ostead, *Proc. Iowa Acad. Sci.*, **71**, 197 (1964).
2. D. E. Worrall, *Org. Syn., Coll. Vol.*, **1**, 413 (1941).
3. See also Farnum and Mostashari, this volume, p. 103.

NORBORNENE DIMER

$$2 \text{ [norbornene]} \xrightarrow[\text{CuBr, Et}_2\text{O}]{h\nu} \text{[norbornene dimer]}$$

Submitted by D. J. TRECKER and R. S. FOOTE*
Checked by T. D. ROBERTS and T. FANNING†

1. Procedure

The apparatus consists of a tubular Pyrex reactor, 15 in. long and 6.9 cm O.D., which contains an internal, concentric, water-jacketed Vycor immersion well, 10 in. long and 4.6 cm O.D. The reactor is equipped with a brine-cooled reflux condenser, a thermocouple well, a septum-capped side arm for sample withdrawal, and a fritted-glass diffuser at the bottom of the reactor for nitrogen ebullition. The light source is a 450-watt medium-pressure Hanovia mercury arc which is suspended in the immersion well.

A solution of 328 g (3.49 moles) of norbornene (Notes 1, 2) in 100 ml of diethyl ether is placed in a loosely stoppered 1-liter flask. Cuprous bromide (1 g) is added, and the mixture is vigorously stirred for 30 minutes. The mixture is filtered into the tubular reactor, brine cooling through the reflux condenser is started, and the solution is continuously sparged with a light flow of oxygen-free nitrogen (Notes 3, 4). Irradiation is then begun; water flow on the lamp jacket is maintained at such a rate that reaction temperature is held at 25–30°. The irradiation is continued for 142 hours (Note 5). During this period the reaction solution is removed from the reactor every 48 hours, treated with a fresh charge of cuprous bromide (0.1 g) and 50 ml of ether to make up for evaporation losses. The Vycor immersion finger is scraped clean of an opaque deposit, and the solution is refiltered into the reactor for continued irradiation.

At the end of the irradiation period the solution is filtered and distilled on a 36-in. spinning-band column (Note 6). After removal of ether at atmospheric pressure, the pressure is gradually lowered to permit distillation of the unreacted norbornene. Finally the dimer (Note 7) is distilled at

---

* Research and Development Department, Union Carbide Corporation, Chemicals and Plastics, South Charleston, West Virginia 25303.
† University of Arkansas, Fayetteville, Arkansas 72701.

50–52° at 1 mm to give 126 g (0.68 mole), 38.3% yield, of fused white material. Recrystallization from 200 ml of absolute ethanol provides 115 g of white crystals (m.p. 65–66°[1,2]) in the first crop (Note 8) and 10 g (m.p. 55–59°) in the second crop (Note 9).

2. Notes

1. The material is obtained from Union Carbide Corp., Chemicals and Plastics, South Charleston, West Virginia.

2. It is essential that the norbornene be flash-distilled (b.p. 92°) and stored under nitrogen before use.

3. Linde H.P. dry nitrogen may be used directly without further purification. Nitrogen of lower purity should be passed through a Fieser train[3] prior to use.

4. If the nitrogen flow is cut off during irradiation and a mineral oil bubbler is used at the exit tube, the brine-cooled reflux condenser is not necessary. Under these conditions, a two-hour nitrogen prepurge should be used.

5. The reaction is followed by gas chromatography, employing a 5 ft × 0.25 in. column of poly(diethanolaminesuccinate) on Chromosorb W at 180°.

6. The use of a spinning-band column is not essential, since the materials being distilled differ widely in boiling points: diethyl ether, 35°; norbornene, 92°; dimer, 50–52° (1 mm).

7. *exo,trans,exo*-Pentacyclo[8.2.1.1^{4,7}.0^{2,9}.0^{3,8}]tetradecane.

8. Analyzed (Note 4) to be 99% *exo,trans,exo*-isomer.[1]

9. Analyzed (Note 4) to be 94% *exo,trans,exo*-isomer and 6% *exo,trans,endo*-isomer.[1]

3. Methods of Preparation

Norbornene dimers have been prepared by catalytic hydrogenation of the corresponding dienes (pentacyclo[8.2.1.1^{4,7}.0^{2,9}.0^{3,8}]tetradeca-5,11-diene),[1] which in turn are prepared by the metal ion-induced dimerization of norbornadiene.[1,4] Mixtures of norbornene dimers are also produced from the photosensitized dimerization of norbornene.[1,5]

4. Merits of the Preparation

The procedure described above has the advantage of yielding a single isomer in high purity from a very simple reaction and workup. The dimer prepared in this manner is suitable for use in the preparation of diadamantane ("congressane").[6]

References

1. D. R. Arnold, D. J. Trecker, and E. B. Whipple, *J. Amer. Chem. Soc.*, **87**, 2596 (1965).
2. D. J. Trecker, J. P. Henry, and J. E. McKeon, *J. Amer. Chem. Soc.*, **87**, 3261 (1965).
3. L. F. Fieser, *Experiments in Organic Chemistry,* 2nd ed., D. C. Heath and Co., Boston, 1941, pp. 395–396.
4. G. N. Schrauzer and S. Eickler, *Ber.*, **95**, 2764 (1962).
5. D. Scharf and F. Korte, *Tetrahedron Lett.*, 821 (1963).
6. C. Cupas, P. von R. Schleyer, and D. J. Trecker, *J. Amer. Chem. Soc.*, **87**, 917 (1964).

PENTACYCLO[4.4.0.0^{2,5}.0^{3,8}.0^{4,7}]DECANE-9,10-DICARBOXYLIC ACID ANHYDRIDE

Submitted by H. G. CUTS, E. N. CAIN, H. WESTBERG, and S. MASAMUNE*
Checked by R. D. MILLER†

1. Procedure

A solution of 10.0 g (0.05 mole) of tricyclo[4.2.2.0^{2,5}]deca-3,7-diene-9,10-dicarboxylic anhydride[1] in 1.1 liters of reagent grade acetone is irradiated (N$_2$ atmosphere, 25°, Hanovia 450-watt type L mercury lamp with a Vycor filter) for 10 hours (Note 1). Removal of the solvent on a rotary evaporator leaves a viscous oil (17 g) which is dissolved in 15 ml of chloroform and chromatographed over 300 g (5.6 cm × 26 cm) of silicic acid (Mallinckrodt, analytical reagent, 100 mesh), using chloroform (0.75% ethanol) as eluent. With this column the chromatographic separation is completed in approximately 3 hours. Fractions 3 through 7 (see Table I)

---

* Department of Chemistry, University of Alberta, Edmonton, Alberta, Canada.
† IBM Research Center, Yorktown Heights, New York 10598.

Table I

Fraction	Volume	Weight	Residue
1	200 ml	—	—
2	175	—	—
3	100	0.19	Semisolid
4	70	0.7	White solid
5	80	3.0	White solid
6	75	1.2	Yellowish solid
7	75	1.6	Yellow solid
8	300	1.2	Yellow oil

are combined to provide 5.8–6.2 g (58–62%) of the pentacyclic anhydride as a waxy yellow solid (Note 2). Colorless needles (4.0–5.0 g), m.p. 127–132° (Note 3) are obtained by dissolving the chromatographic residue in 20 ml of hot benzene, adding 40 ml of hot cyclohexane, and finally concentrating the solution (at atmospheric pressure) to approximately 15 ml.

2. Notes

1. The cyclization is essentially complete after 8 hours of irradiation. The nmr spectrum shows only small amounts of olefinic compounds to be present and exhibits virtually no change after 8 hours.

2. This residue contains as high as 10% of tricyclo[4.2.2.0^{2,5}]dec-7-ene-9,10-dicarboxylic anhydride which must result from photoreduction of the cyclobutene double bond in tricyclo[4.2.2.0^{2,5}]deca-3,7-diene-9,10-dicarboxylic anhydride.

3. The melting point and yield depend upon the quality of starting material and technique of recrystallization. An analytically pure sample melts slightly higher than that recorded herein.

3. Methods and Merits of the Preparation

This is the only method that has been reported for the preparation of pentacyclodecane-9,10-dicarboxylic anhydride. It will remain the best one because of the simplicity of the experiments and the availability of the starting material.

References

1. W. Reppe, O. Schlichting, K. Klager, and T. Toepel, *Ann. Chem.*, **560**, 1 (1948).

anti-PENTACYCLO[5.3.0.0^{2,5}.0^{3,9}.0^{4,8}]DECAN-6-OL

anti-1,3-Bishomocubanol

Submitted by WENDELL L. DILLING and CHARLES E. REINEKE*
Checked by T. D. ROBERTS†

1. Procedure

The ultraviolet radiation source is a 450-watt Hanovia medium-pressure mercury arc lamp, type 679A. The lamp is placed inside a tubular 9700 Corex glass filter ($\lambda > \sim$260 mµ) in a water-cooled quartz well (Note 1). The well is inserted into a cylindrical reaction vessel which has a 60/50 standard taper joint on top. The reaction vessel is constructed so that 150 ml of solution will cover the well to a level above the window of the lamp when in place (Note 2). The reaction vessel also contains a gas inlet in the bottom, a side arm near the top for removing samples, and a neck for a condenser with a gas outlet tube on top.

A solution of 10.0 g (0.068 mole) of *endo,anti*-tricyclo[5.2.1.0^{2,6}]deca-4,8-dien-3-ol (α-1-hydroxydicyclopentadiene) (Note 3) in 150 ml of distilled acetone (Note 4) is placed in the reaction vessel, and a stream of purified nitrogen is bubbled through the solution for *ca.* 30 minutes to remove dissolved oxygen. The solution is then irradiated under nitrogen at room temperature (Note 5). The course of the reaction may be followed by periodically removing samples for gas chromatographic analysis. This analysis can be carried out using a 10 ft × ¼ in. 20% Apiezon L on 60/80 mesh Chromosorb WAW column operated isothermally at 225° and a helium flow rate of 40 cc/min. Under these conditions, acetone is eluted very quickly and followed by the starting dienol (retention time 8.2 minutes) and the product alcohol (retention time 9.7 minutes). After 6–7 hours of

---

* Edgar C. Britton Research Laboratory, The Dow Chemical Company, Midland, Michigan 48640.
† University of Arkansas, Fayetteville, Arkansas 72701.

irradiation the conversion of the starting dienol into the product alcohol is essentially complete (Note 6).

The acetone is removed from the reaction mixture under vacuum to give 12.8 g of viscous yellow oil containing some crystals. This oil is vacuum-sublimed twice at 0.5 mm and a bath temperature of 110° to give 5.1 g of soft white crystals (Note 7). Recrystallization once from hexane gives 3.1 g (31%) of product, m.p. 145°–155°. Additional purification may be achieved by further recrystallization from hexane, m.p. 164°–166° (Note 8).

2. Notes

1. The lamp, Corex glass filter, and quartz well are available from the Hanovia Lamp Division, Engelhard Hanovia, Inc., Newark, New Jersey.

2. The size of the reaction vessel may be varied in accordance with the amount of solution to be irradiated.

3. The hydroxydicyclopentadiene is conveniently prepared in 50–70% yield by the selenium dioxide oxidation of commercially available *endo*-dicyclopentadiene as described by Woodward and Katz.[1] Dienol distilling at 67–74° (0.5 mm) was used in the photoreaction. No further purification was carried out.

4. The acetone was not purified beyond distillation.

5. No cooling is required other than cold water running through the lamp well. The solution may warm slightly above room temperature.

6. The checker used a Hanovia 200-watt lamp and a more dilute solution (8.1 g in 400 ml). The irradiation time was 24 hours.

7. A water-cooled cold finger in the sublimation apparatus is sufficient to collect the product.

8. A melting point of 171–172° has also been reported for this product.[2]

3. Methods of Preparation

Other than by the method described, *anti*-1,3-bishomocubanol has been prepared by acetolysis of pentacyclo[4.4.0.0^{2,5}.0^{3,8}.0^{4,7}]dec-9-yl methanesulfonate (1,1-bishomocubyl mesylate) followed by hyrolysis,[3] the sodium borohydride[3] or lithium aluminum hydride[2,4,5] reduction of pentacyclo-[5.3.0.0^{2,5}.0^{3,9}.0^{4,8}]decan-6-one (1,3-bishomocubanone), and irradiation of *endo,syn*-tricyclo[5.2.1.0^{2,6}]deca-3,8-dien-10-ol (*syn*-8-hydroxydicyclopentadiene[1]) in acetone solution.[2,5]

4. Merits of the Preparation

The present procedure has the advantage of using a starting material which is prepared in one step in good yield from inexpensive commercially available starting material. The reaction and purification are relatively simple procedures to carry out. The nonphotochemical routes described in Section 3 give the *anti*-alcohol as a minor product (20% or less) which is virtually impossible to separate from isomeric alcohols.[4,5] This synthesis can be adapted to a number of other derivatives of 1,3-bishomocubanes from the appropriately substituted *endo*-dicyclopentadienes.

References

1. R. B. Woodward and T. J. Katz, *Tetrahedron*, **5**, 70 (1959).
2. R. C. Cookson, J. Hudec, and R. O. Williams, *J. Chem. Soc., C*, 1382 (1967).
3. W. G. Dauben and D. L. Whalen, *J. Amer. Chem. Soc.*, **88**, 4739 (1966).
4. W. L. Dilling and M. L. Dilling, *Tetrahedron*, **23**, 1255 (1967).
5. W. L. Dilling and C. E. Reineke, *Tetrahedron Lett.*, 2547 (1967).

2-PHENYLBICYCLO[1.1.1]PENTANOL-2

Submitted by ALBERT PADWA and EDWARD ALEXANDER*
Checked by F. T. BOND†

1. Procedure

The irradiation flask used is designed to fit around a Hanovia immersion well. The flask is fitted with a side arm to a condenser, a nitrogen inlet tube, and an outlet fitted with polyethylene tubing to allow for removal of aliquots. The solutions are stirred magnetically. A Hanovia 550-watt high-pressure mercury arc inside a Pyrex filter sleeve is used as the light source.

---

* State University of New York, Buffalo, New York 14214.
† University of California, La Jolla, California 92037.

The photolysis is carried out by dissolving 5.0 g of cyclobutyl phenyl ketone (Note 1) in 1 liter of freshly distilled benzene. The solution is purged with nitrogen for 45 minutes prior to photolysis, and positive pressure maintained during the run (Note 2). The photolysis is monitored by removing aliquots for glpc analysis (Notes 3, 4). The photolysis is at least 90% complete after 60 hours; however, the photolysis is allowed to go a total of 72 hours. The solvent is evaporated on a rotary evaporator under reduced pressure at *ca.* 30°. The residues from three successive photolyses are combined (Note 5) and distilled through an 18-in. spinning-band column to yield 5.65 g of a colorless liquid, b.p. 65–68° (0.2 mm). This material solidifies on standing and is placed on a porous plate to remove the residual oils. The product may be purified by sublimation at 54° (0.1 mm). The yield of 2-phenylbicyclo[1.1.1]pentanol-2, m.p. 64–65°, is 3.75 g (25%).

2. Notes

1. Cyclobutyl phenyl ketone (Aldrich Chemical Co.) was used from a freshly opened bottle without further purification.

2. The nitrogen atmosphere was purified by the vanadous ion method of Meites.[1]

3. Gas-liquid partition chromatographic analyses were carried out using an F and M model 5750 instrument equipped with a 5 ft × $\frac{1}{8}$ in. 5% DEGS on chromosorb W column at a flow rate of 60 cc/min and at a temperature of 120°.

4. In addition to starting material (retention time, 5.6 minutes) and phenylbicyclo[1.1.1]pentanol (8.0 minutes), peaks corresponding to 1-phenyl-4-penten-1-one (4.5 minutes) and cyclobutyl phenyl carbinol (11.0 minutes) also appeared in the vapor phase chromatogram.[2]

5. Maximum yields of phenylbicyclo[1.1.1]pentanol were realized when the irradiation was carried out on 5.0-g batches of cyclobutyl phenyl ketone.

3. Methods of Preparation

Related bicyclo[1.1.1]pentanes have been prepared by the reaction of 3-(bromomethyl)cyclobutyl bromide with lithium amalgam,[3] from the gas phase irradiation of bicyclo[2.1.1]hexan-2-one,[4] and from the Hg($3p_1$)-sensitized isomerization of substituted 1,4-pentadienes.[5]

The present procedure has the advantage of using commercially available starting material and also provides for sufficient quantities of functionalized material.

References

1. L. Meites and T. Meites, *Anal. Chem.,* **20,** 984 (1948).
2. A. Padwa and E. Alexander, *J. Amer. Chem. Soc.,* **89,** 6376 (1967).
3. K. B. Wiberg and D. S. Connor, *J. Amer. Chem. Soc.,* **88,** 4437 (1966).
4. J. Meinwald, W. Szkrybalo, and D. R. Dimmer, *Tetrahedron Lett.,* 731 (1967).
5. R. Srinivasan and K. H. Carlough, *J. Amer. Chem. Soc.,* **89,** 4932 (1967).

5-PHENYLPHENANTHRIDINE-6[5H]-ONE

Submitted by EDWARD C. TAYLOR and GAVIN G. SPENCE*
Checked by T. D. ROBERTS and M. CONDIT†

1. Procedure

The apparatus consists of a Pyrex tube, 31.0 cm long and 5.1 cm O.D., closed at one end and fitted at the open end with a 24/40 standard taper joint. The source of ultraviolet (3000 Å) radiation is a Rayonet RPR-100 reactor (Note 1).

The reaction vessel is filled with a solution of 4.00 g (0.0148 mole) of 6-phenylphenanthridine N-oxide (Note 2) in 400 ml of absolute ethanol (Note 3). The vessel is fitted with a reflux condenser, and the solution is irradiated for 20 hours (Notes 4, 5). The mixture of white solid and yellow solution formed is concentrated to a volume of about 100 ml. The precipitated product is collected and washed with a small amount of ethanol. The yield of 5-phenylphenanthridine-6[5H]-one is 3.51 g (88%) of pale yellow crystals (m.p. 223.5–225.5°) (Note 6). The product can be further purified by passing it through a column of Florisil (Note 7) with chloroform as eluent (Note 8). In a typical run, 3.00 g of crude product is passed through a 100-g column of Florisil. The product recovered consists of

---

* Princeton University, Princeton, New Jersey 08540.
† University of Arkansas, Fayetteville, Arkansas 72701.

2.57 g (86% recovery) of a white solid (m.p. 227.8°, 220.0°, 0.2° per minute) (Note 9) that is homogeneous to thin-layer chromatography.

2. Notes

1. Supplied by Southern New England Ultraviolet Company, Middletown, Connecticut.

2. The starting material is prepared in the following way: a solution of 50.0 g (0.25 mole) of 2-aminobiphenyl (Aldrich Chemical Company, Inc., cat. No. A 4240-9) in 250 ml of pyridine is cooled to −5° in an ice-salt bath. The solution is stirred magnetically, and the temperature is maintained below 0° as 35.1 g (0.25 mole) of benzoyl chloride is slowly added dropwise. The solution is allowed to warm to room temperature and is then stirred for 30 minutes. The resulting mixture is poured into 1 liter of iced dilute hydrochloric acid. The product is extracted with ether, and the residue from the ether extracts is recrystallized from benzene-petroleum ether. The yield of 2-benzamidobiphenyl is 58.3 g (85.3%) of a white solid (m.p. 80–85°).

A solution of 58.3 g (0.213 mole) of 2-benzamidobiphenyl in 240 g of polyphosphoric acid (Matheson, Coleman and Bell, cat. No. 8096) is stirred magnetically and heated at 120–130° for 6 hours. The solution is then poured into water, and the resulting mixture is neutralized with dilute sodium hydroxide solution. The precipitated 6-phenylphenanthridine is collected and recrystallized from ethanol-hexane. The yield is 37.9 g (69.6%) of a pale yellow solid (m.p. 106–109°).

To a solution of 30.7 g (0.12 mole) of 6-phenylphenanthridine in 200 ml of chloroform is added 24.3 g (0.12 mole) of 85% *m*-chloroperbenzoic acid (FMC Corporation). The resulting solution is allowed to stand overnight in the dark at room temperature. The precipitated *m*-chlorobenzoic acid is removed, and the chloroform solution is washed with dilute sodium hydroxide solution to remove the remaining acids. The chloroform solution is then evaporated to dryness. The 6-phenylphenanthridine N-oxide obtained is recrystallized from ethanol. The yield is 11.7 g (35.8%) of yellow crystals (m.p. 223–224°). TLC of this product shows that it is very slightly contaminated with unoxidized 6-phenylphenanthridine, which can be removed by washing a methylene chloride solution of the crude N-oxide with dilute hydrochloric acid.

The pure N-oxide is very sensitive to light. It has been shown that the solid undergoes a photochemical rearrangement to some extent on prolonged exposure to normal laboratory light. Thus it is necessary to keep the solid and solutions in the dark as much as possible.

3. The reaction is very dependent on the solvent used. In aprotic solvents, products other than the phenanthridone are isolated. For example, in benzene solution, N-benzoylcarbazole, 2-phenyldibenzo[d,f]-1,3-oxazepine, and 6-phenylphenanthridine are formed in addition to 5-phenylphenanthridine-6[5H]-one.

4. After irradiation for 20 hours, TLC indicates that some starting material is still present. However, further irradiation is undesirable because most of the incident light is then absorbed by the product. It has been shown that the product decomposes slightly on prolonged irradiation.

5. Some of the product, 5-phenylphenanthridine-6[5H]-one, separates from the solution during the irradiation. However, it falls to the bottom of the reaction vessel and thus does not interfere with absorption of light by the solution.

6. The filtrate contains more of the desired product, but it also contains a higher percentage of unreacted starting material. By careful chromatography of the entire irradiation mixture, a yield as high as 96% (based on recovered starting material) of pure 5-phenylphenanthridine-6[5H]-one has been achieved.

7. Florisil can be obtained from the Floridin Company, Hancock, West Virginia (100/200 mesh is used).

8. Recrystallization of the crude product from ethanol-chloroform yields 2.45 g (61.3%) of 5-phenylphenanthridine-6[5H]-one. The recrystallized product has the correct melting point (226–228°) and spectral properties, but its TLC still shows a spot due to unreacted starting material.

9. The first melting point was recorded on a Mettler FP-1 apparatus. The second number indicates the temperature at the start of heating, and the third figure indicates the rate of heating of the sample.

3. Methods of Preparation

5-Phenylphenanthridine-6[5H]-one has been prepared by a Chapman rearrangement on 6-phenoxyphenanthridine[1] and thermal or catalytic decomposition of diazonium salts of N-o-aminobenzoyldiphenylamine.[1,2]

4. Merits of the Preparation

The present procedure has the advantage of using a starting material available from a three-step synthesis. The nature of the synthesis allows the preparation of a large variety of substituted phenanthridine N-oxides. Each step of the synthesis is simple and produces the desired products in reasonably high yield. This is a reaction involving migration of a substituent from carbon to nitrogen, and it occurs under extremely mild conditions.

References

1. D. H. Hey and T. M. Moynehan, *J. Chem. Soc.*, 1563 (1959).
2. T. M. Moynehan and D. H. Hey, *Proc. Chem. Soc.* (London) 209 (1957).

PHOTOCHEMICAL WOLFF REARRANGEMENT

exo-2,3-Benzonorcaradiene-7-acetic acid (III)

Submitted by ANDREW S. KENDE and ZEEV GOLDSCHMIDT*

1. Procedure

A. Diazomethyl ketone (II). To a solution of 5.00 g (0.0269 mole) of the *exo*-2,3-benzonorcaradiene-7-carboxylic acid (I) (Note 1) in 100 ml of dry methylene chloride (Note 2), stirred magnetically in a flask protected from moisture, is added a solution of 4.00 g (0.0336 mole) of thionyl chloride in the same solvent, followed by 0.20 g (0.0027 mole) of dimenthylformamide as catalyst (Note 3). The mixture is stirred at room temperature for 5 hours. Complete conversion into the acid chloride is monitored by infrared spectroscopy near 1795 and 1710 cm^{-1}. Removal of both solvent and excess thionyl chloride at reduced pressure on a rotary

* University of Rochester, Rochester, New York 14627.

evaporator below 35° gives the crude, crystalline acid chloride which is used without further treatment.

A solution of this chloride in 100 ml of dry ether is added to an ice-cold solution of freshly prepared diazomethane (from 15 g crude N-nitrosomethylurea) (Note 4) in 150 ml of ether. The reaction mixture is stirred magnetically and allowed to come to room temperature over the course of 3 hours, filtered through an ether-moistened cotton plug to remove insolubles, then stripped of solvent on a rotary evaporator at 25° at reduced pressure in the hood. The solid yellow diazoketone (II) which is formed is once recrystallized from ether to give 4.51 g (87%), m.p. 108–110°. This product exhibits the characteristic diazoketone infrared spectrum containing bands at 2080 and 1620 cm^{-1}, and gives nmr data in accord with the indicated structure.

B. Exo-*2,3-Benzonorcaradiene-7-acetic acid* (*III*). To a solution of 2.16 g (0.0103 m) of the diazoketone (II) in 80 ml of acetonitrile is added 20 ml of water. The solution is placed in the inner well of a water-jacketed photolysis vessel (Pyrex) and irradiated at 20–25° using a medium-pressure Hanovia utility lamp (Note 5) as an external light source. After 20 hours the infrared spectrum shows virtual disappearance of the diazoketone bands. The reaction mixture is concentrated on a rotary evaporator at reduced pressure, and the resulting syrup partitioned between saturated sodium bicarbonate solution and ether. The aqueous layer is acidified with concentrated hydrochloric acid, and extracted at strongly acidic pH with three portions (total 200 ml) of ether, the ether extracts washed with a small volume of water, then dried over magnesium sulfate. Removal of solvent under reduced pressure and recrystallization of the acid from petroleum ehter (b.p. 30–60°) gives 0.83 g (40%) of pure *exo* acid (III), m.p. 92–93° (Notes 6, 7).

2. Notes

1. Preparation of the starting material from the reaction of naphthalene and ethyl diazoacetate is given by Huisgen and Juppe. [1]
2. The methylene chloride was dried over molecular sieves for 48 hours before use.
3. For this modification, see Bosshard and co-workers.[2]
4. For this procedure, see Arndt.[3]
5. The lamp was model No. 30620 with a nominal 140-watt rating.
6. The yield of *exo* acid (III) has a theoretical maximum of 65%, since under the reaction conditions a slow equilibration between the *exo* and *endo* isomers takes place, and 65% is the equilibrium percentage of *exo*

form. On the basis of this fact, the observed 0.83 g corresponds to a corrected yield of 62%.

7. For physical data on acid (III), see Kende and co-workers.[4]

References

1. R. Huisgen and G. Juppe, *Chem. Ber.,* **94,** 2332 (1961).
2. H. H. Bosshard, R. Mory, M. Schmid, and Hch. Zollinger, *Helv. Chim. Acta,* **42,** 1653 (1959).
3. F. Arndt, *Org. Syn. Coll. Vol.,* **2,** 165 (1943).
4. A. S. Kende, Z. Goldschmidt, and P. T. Izzo, *J. Amer. Chem. Soc.,* **91,** 6858 (1969)

TETRACYCLO[3.2.0.0^{2,7}.0^{4,6}]HEPTANE-2,3-DICARBOXYLIC ACID

Submitted by STANLEY J. CRISTOL and ROBERT L. SNELL*
Checked by T. D. ROBERTS and T. FANNING†

1. Procedure

A 500-ml Vycor Erlenmeyer flask containing 2.01 g (0.011 mole) (Note 1) of bicyclo[2.2.1]hepta-2,5-diene-2,3-dicarboxylic acid (Note 2) and 200 ml of anhydrous ether (Note 3) is fitted with a magnetic stirrer and a reflux condenser (Note 4) vented out a window by a tube and drying tower (Note 5). The apparatus is placed in a fume hood (Note 6). The stirred solution is irradiated for 11 hours by a 360-watt General Electric Uviarc (UA-3) quartz tube mercury vapor lamp set 12 in. from the center of the flask (Note 7). The white precipitate, 1.63 g, is removed by filtration and washed with anhydrous ether. The combined filtrates are taken almost to dryness on a steam bath (Note 8). The slightly yellowish residue is dissolved in acetone (Note 9). The solution is concentrated to

* University of Colorado, Boulder, Colorado 80302.
† University of Arkansas, Fayetteville, Arkansas 72701.

8–10 ml, cooled, and filtered. The precipitate, 0.19 g, is washed with acetone.

Thus the total recovery of tetracyclo[3.2.0.0^{2,7}.0^{4,6}]heptane-2,3-dicarboxylic acid is 1.82 g, 90% of theory (Note 10). The melting-point behavior of the product is complex, but the solid usually shows an abrupt change in appearance in the range 170–175° and always shows a gradual foaming beginning in the range 215–220° (Note 11).

The photo product is characterized by ir peaks (low concentration in KBr pellet) at 3081, 1675, and 1620 cm^{-1} and by an ultraviolet maximum at 233 mμ (log ε 3.41).

2. Notes

1. The concentration is not critical.
2. Prepared by the method of Alder and Brochhagen.[1]

Commercially obtainable monopotassium salt of acetylene dicarboxylic acid was treated as described in *Organic Syntheses*,[2] but the aqueous solution was filtered before ether extraction. The combined ethereal extracts, after washing, were dried over anhydrous magnesium sulfate and concentrated. Two alternatives were followed. In one, the acid, m.p. 189–190°, was precipitated by addition of dry reagent grade benzene. In the other, which is probably better, the solution, in a loosely corked flask, was placed in a Dry Ice chest both for cooling and to saturate the ether with carbon dioxide. The acid obtained in this way had m.p. 179–180°.

Dicyclopentadiene was pyrolyzed to cyclopentadiene, which was redistilled and immediately used in the adduction.

3. The photo product is known to react with water contained in various solvents. Thus solvents must be dry.
4. Either tightly fitted corks or ground-glass joints may be used.
5. This system of venting was used primarily to isolate the flask contents from the ozone produced by the lamp.
6. The lamp used here generates large quantities of ozone.
7. The original synthesis used a G.E. AH-4 lamp with the outer envelope removed to expose the quartz tube.
8. *Caution!* Ether peroxides may be present in quantities ranging from insignificant to dangerous. Anhydrous ether should be used from freshly opened cans. This problem can be avoided by accepting slightly lower yields and discarding the mother liquors.
9. Because the commercial reagent acetone on hand contained considerable quantities of condensation products, acetone distilled from permanganate was used to avoid a contaminated product.

10. A simultaneous companion experiment using a Pyrex flask gave a total photo product of 1.75 g from 2.00 g of starting acid. It has also been shown that 16 hours of irradiation in sunlight gives essentially a quantitative yield whether in a Pyrex or Vycor flask. The workup and results are as described here.

11. The product, as obtained directly from ether, is normally white. There are no impurities detectable by uv, ir, or nmr spectroscopy. If the product is not white, it can be recrystallized from acetone. However, recrystallization of the product does not lead to a predictable melting behavior. Some samples seem to "jump" in the range 170–175° and fall back to a dry powder, others fall back to an amorphous mass, and others appear to do nothing at all. No further effects are noted until the range 215–220°, where a foam slowly forms. This foam may or may not show evidence of liquid formation in the region of 240°.

3. Methods of Preparation

Tetracyclo[3.2.0.0^{2,7}.0^{4,6}]heptane-2,3-dicarboxylic acid has been prepared only by this method.[3] Edman and Simmons have noted (ref. 4, p. 3814) that benzene does not serve as a solvent in this preparation.[4]

References

1. K. Alder and F. Brochhagen, *Chem. Ber.*, **87**, 167 (1954).
2. *Org. Syn. Coll.* **2**, 10 (1943).
3. S. J. Cristol and R. L. Snell, *J. Amer. Chem. Soc.*, **76**, 5000 (1954); S. J. Cristol and R. L. Snell, *ibid.*, **80**, 1950 (1958).
4. J. R. Edman and H. E. Simmons, *J. Org. Chem.*, **33**, 3808 (1968).

TETRACYCLO[3.2.0.0^{2,7}.0^{4,6}]HEPTANE (QUADRICYCLENE)

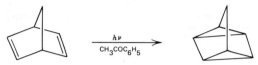

Submitted by F. I. SONNTAG and R. SRINIVASAN*
Checked by J. J. MAHER and E. J. MORICONI†

1. Procedure

The apparatus consists of a cylindrical pyrex vessel equipped with a 60/50 ground joint for accommodating the light source, a side arm for holding a reflux condenser, and a magnetic stirring bar. A Hanovia 450-watt high-pressure mercury arc is used in conjunction with a Pyrex, double-walled cooling well (Note 1). External cooling is provided by an ice bath.

The reaction vessel is charged with a solution of 25 ml (22.8 g, 0.25 mole) of bicyclo[2.2.1]hepta-2,5-diene (norbornadiene) in 1280 ml of petroleum ether (38–53°) (Note 2), and 50 ml (51.5 g) of acetophenone (Note 3). Nitrogen is bubbled through the stirred solution for 5 minutes, the irradiation is then started and continued for 5 hours. The temperature of the reaction mixture is maintained below 15° (Note 4).

The irradiated mixture is filtered into a 2.0-liter, round-bottomed flask (Note 5) and submitted to distillation through a 60-cm "Nester-Faust" Teflon spinning-band column, using a water bath. The cuts taken are given in Table I. Cut 4 consists of 93% quadricyclene (Note 6) and 7% norbornadiene, corresponding to a total yield of 57% based on pure quadricyclene. If required, the product can be purified further by vapor phase chromatography (Note 7).

2. Notes

1. Lamp cat. No. 679-36. An immersion well similar to the one used, but made of quartz, is supplied by Hanovia, cat. No. 19434.
2. A small amount of a white precipitate was formed when norbornadiene was dissolved in petroleum ether. The filtered solution was used for irradiation.

---

* IBM Research Center, Yorktown Heights, New York 10598.
† Fordham University, The Bronx, New York 10400.

Table I

Cut	Water bath temperature	Pressure	Reflux ratio	Head temperature	Time, hours	Distillate volume, ml
1	47–53°	1 atm	1:1	33–39°	3	500 ml
2	53–69°	1 atm	1:1–2:1	40–54°	3	500 ml

(The solution is transferred to a 500-ml flask which is heated directly and the distillation is continued.)

3	35–77°	210 mm	5:1–50:1	20–27°	$1\tfrac{3}{4}$	50 ml
4	112–175°	515 mm	5:1	67°		18 ml (14 g)

3. All the reagents used were commercial materials: bicyclo[2.2.1]hepta-2,5-diene, Aldrich Chemical Co.; petroleum ether and acetophenone, Fisher Chemical Co.

4. When the temperature of the reaction mixture was allowed to rise above 15°, the conversion into product was reduced.

5. The filtration is required in order to remove the white solid which is formed in small amounts during the irradiation.

6. It is desirable not to heat the quadricyclene to high temperatures in order to avoid its reversal to norbornadiene.

7. An 8-ft silicone (UCON-550X) column was used at 60°. Under these conditions the relative retention times of norbornadiene and quadricyclene are 1.0 and 2.2, respectively.

3. Methods of Preparation

Quadricyclene has been prepared by the photoisomerization of bicyclo-[2.2.1]hepta-2,5-diene by direct irradiation[1] and on sensitization.[2] The procedure described here is the latter.

References

1. W. G. Dauben and R. L. Cargill, *Tetrahedron*, **15**, 197 (1961).
2. G. S. Hammond, N. J. Turro, and A. Fischer, *J. Amer. Chem. Soc.*, **83**, 4674 (1961).

1,3,5,7-TETRAMETHYLCYCLOOCTATETRAENE

Submitted by P. DE MAYO, C. L. McINTOSH, and R. W. YIP*
Checked by J. K. CRANDALL and R. J. SEIDEWAND†

1. Procedure

The apparatus consists of a water-cooled immersion well containing a Pyrex filter and a 450-watt Hanovia medium-pressure mercury arc lamp, Model 79A36. A solution of 20.0 g of 4,6-dimethyl-α-pyrone[1] (Note 1) in 15 ml of dry benzene (Note 2) is placed in three 9-mm Pyrex tubes and sealed with rubber stoppers after degassing with oxygen-free nitrogen for 30 minutes. The tubes are strapped to the outside of the immersion well, the apparatus wrapped with aluminum foil and immersed in a 4-liter beaker of water.

The irradiation is followed by monitoring the disappearance of the infrared absorption band at 1560 cm^{-1} of the α-pyrone (Note 3); approximately 100 hours is required. The total irradiation mixture is then transferred to a 50-ml, round-bottomed flask with a side arm and a standard vacuum distillation apparatus assembly.

After the solvent has been removed, the residue is pyrolyzed with a free flame for 15 minutes (Note 4), taking care that none of the material distills over. Vacuum is then applied and the cyclooctatetraene distilled at 85–95° (15 mm). When the distillation is completed, the vacuum is released, the pot material pyrolyzed for an additional 10 minutes, and distillation continued to give a total of 2.21–2.65 g of 1,3,5,7-tetramethylcyclooctatetraene, m.p. 63–67°.

The cyclooctatetraene is purified by passage down a column of 25 g of silicic acid, using 35–60° light petroleum as eluent, to give 1.98–2.17 g (16–18%), m.p. 69.5–70°; $v_{max}^{CCl_4}$ 1650 cm^{-1}; δ (CCl$_4$) 1.68 (12 H, s), 5.35 (4 H, s); λ_{max}^{EtOH} 285 mμ (ε 470).

* University of Western Ontario, London, Ontario, Canada.
† Indiana University, Bloomington, Indiana 47401.

Further distillation of the pot material gives 5.4–7.0 g of 4,6-dimethyl-α-pyrone which, when recrystallized, has m.p. 48–50°, 5.0–6.2 g.

The isolated yield of 1,3,5,7-tetramethylcyclooctatetraene is 16–18%, but the yield after correction for recovered 4,6-dimethyl-α-pyrone is 21–25%.

2. Notes

1. The 4,6-dimethyl-α-pyrone was prepared as described by Wiley[1] and recrystallized from ethyl acetate before use, m.p. 49–50° (lit. m.p. 50–51°), λ_{max}^{EtOH} 295 mµ (ε 6,200).

2. Although the pyrone will not initially dissolve completely in this volume of benzene, this does not appear to have a detrimental effect on the reaction.

3. It may be necessary to filter the reaction mixture at some point during the reaction and transfer the mother liquors to new tubes, since light-absorbing material has a tendency to deposit on the sides of the tubes. It is usually possible, however, to get greater than 90% reaction of the α-pyrone.

4. A Wood's metal bath at 230–35° was also used, but with no apparent advantage.

Reference

1. N. R. Smith and R. H. Wiley, *Org. Syn.*, **32**, 57, 76 (1952).

TRICYCLO[3.3.0.0²,⁶]OCTANE

Submitted by R. SRINIVASAN*
Checked by J. PICONE*

1. Procedure

The irradiation is carried out in a cylindrical quartz container (8 cm O.D., 25 cm long), of 1.5 liters capacity (Note 1), the top end of which is fitted with a 24/40 ground joint. A solution of 1,5-cyclooctadiene (14 g, 0.13 mole) (Note 2) in 1 liter of ether is placed in the vessel and 100 mg of freshly prepared cuprous chloride (Note 3) is added. The vessel is closed with a glass stopper which is held in place with a spring, and the solution is stirred magnetically for 30 minutes (Note 4). The vessel is placed in a Rayonet RPR-100 reactor (Note 5) fitted with 253.7-nm lamps and the fan is turned on.

After 24 hours the dark solution is filtered, more cuprous chloride (100 mg) is added with stirring, and the irradiation is continued (Note 6). The filtration and resaturation with cuprous chloride is repeated after 48 hours and again after 200 hours. The total irradiation time required is 360 hours (Note 7).

The photolyzed solution is distilled on a spinning-band distillation column. After the ether is removed, the temperature rises rapidly to 120°. The distillate (10.9 g) is collected at 80–120°. The residue (∼3.5 g) is discarded.

The distillate, which is tricyclo[3.3.0.0^{2,6}]octane of about 80% purity, is refluxed with an aqueous solution of potassium permanganate (20 g in 200 ml) for 1 hour. The solution is then steam-distilled until no oily drops are seen to come over. The distillate is extracted with pentane (two 30-ml portions) and the aqueous layer is discarded. The combined pentane extracts are dried over calcium chloride and distilled on a micro spinning-band column. After the pentane has been removed, tricyclo[3.3.0.0^{2,6}]octane (6.0 g, 43%), b.p.$_{740}$127.5°, n^{23}D 1.4709, distills over. The material is 96–97% pure.

* IBM Research Center, Yorktown Heights, New York 10598.

The nmr spectrum of tricyclo[3.3.0.0^{2,6}]octane consists of two singlets at τ 8.18 and 8.27 in the ratio 1:2.

2. Notes

1. Neither the design nor the dimensions of the container are critical.
2. Any commercial sample of 1,5-cyclootadiene of better than 95% purity is suitable. It is fractionally distilled before use.
3. The procedure given by Vogel[1] is satisfactory.
4. The stopper, when held in place with a spring acts as a safety valve if the pressure builds up inside the vessel at this point or later during the irradiation. Alternatively, the vessel can be closed with a reflux condenser and a drying tube. The evaporation of the ether during irradiation is then appreciable and should be compensated. The solubility of cuprous chloride in the solution is small.
5. Supplied by Southern New England Ultraviolet Co., Middletown, Connecticut. *Caution!* Any other source of 253.7 nm radiation can be used, but lamps that run very hot can create a fire hazard because of the large quantity of ether that is used.
6. The container should be cleaned before the solution is poured into it again.
7. The progress of the reaction can be followed by vpc analysis. On a 2-meter UCON-550x column at 90°, the retention times of the product and 1,5-cyclooctadiene are 18 and 48 minutes, respectively. At the end of the irradiation the ratio of the product to diene is 20:1.

3. Methods of Preparation

Tricyclo[3.3.0.0^{2,6}]octane has been prepared by the photoisomerization of 1,5-cyclooctadiene with either mercury ($^3p_1$) atoms in the vapor phase as the sensitizer[2] or with cuprous chloride as a catalyst in solution.[3,4] The procedure described here is the latter.[3] The workup is according to that in Reference 4.

References

1. A. I. Vogel, *Practical Organic Chemistry,* John Wiley and Sons, New York, 1962, p. 191.
2. R. Srinivasan, *J. Amer. Chem. Soc.,* **85,** 819 (1963).
3. R. Srinivasan, *J. Amer. Chem. Soc.,* **86,** 3318 (1964).
4. J. Meinwald and B. E. Kaplan, *J. Amer. Chem. Soc.,* **89,** 2611 (1967).

α-TRUXILLIC ACID*

Cyclobutane-*trans*-1,3-dicarboxylic acid, *trans*-2,4-diphenyl-

Submitted by DONALD G. FARNUM and ABDOL J. MOSTASHARI†
Checked by L. WILLIAMS and R. SRINIVASAN‡

1. Procedure

The photolysis of *trans*-cinnamic acid (Note 1) is carried out in a 5-liter, two-necked standard taper (55/60 and 24/40) round-bottomed flask. A mechanical stirrer fitted with a 1–1.5 cm Teflon paddle is inserted through the smaller neck and a Hanovia-type Pyrex photolysis insert is placed in the larger neck when needed. The acid (145 g, 0.98 m) is suspended in 500 ml of water in the flask with vigorous mechanical stirring. An additional 4 liters of water is then introduced along with a football-type magnetic stirrer. The continuously stirred suspension is then photolyzed for 5 days by means of a 550-watt Hanovia lamp placed in the Pyrex insert (Note 2). The solid is filtered and air-dried. The dry white solid is washed with ether several times, then placed in a 1-liter beaker, triturated with 500 ml of ether, and filtered once more. The crude solid (81.5 g, 56%) (Note 3), m.p. 280–285°, can be purified further by recrystallization from 95% ethanol using a little decolorizing charcoal. In this way 60–65 g of pure α-truxillic acid, m.p. 290–292° (lit.[1] 283–284°) can be obtained.

Large quantities of unreacted *trans*-cinnamic acid can be obtained from the ether washings by evaporation and recrystallization of the resultant yellow solid.

2. Notes

1. Practical *trans*-cinnamic acid, m.p. 132–134°, was used. Best results were obtained with material supplied by Fisher Scientific, although samples from Matheson and Malinckrodt are satisfactory.

---

* This work was supported by the National Science Foundation under Grant GP 10734.
† Michigan State University, East Lansing, Michigan 48823.
‡ IBM Research Center, Yorktown Heights, New York 10598.

2. The checkers used four Rayonet RPR-208 modules with 3000 Å lamps placed externally. The reaction rate was fourfold slower.

3. Yields of crude product varied from 45–65%. Reasons for the variation were not determined, although efficiency of stirring is thought to be important.

3. Methods and Merits

The procedure described is a modification of the method of White and Dunathan.[1] It has the advantage of permitting large-scale preparations in good yield without unwieldy apparatus. The solid state photodimerization of α-*trans*-cinnamic acid[2] proceeds in moderate yield but is inefficient for large-scale preparations.

References

1. E. H. White and H. C. Dunathan, *J. Amer. Chem. Soc.*, **78**, 6005 (1956). This preparation gave 28.3 g of product in 9.5% yield after 7 days of irradiation.
2. L. H. Klemm, K. W. Gopinath, T. J. Dooley, and C. E. Klopfenstein, *J. Org. Chem.*, **31**, 3003 (1966). This preparation gives 4.5 g of product in 33% yield after 8 days of irradiation.

INDEX

2-Acetonaphthone, 30, 49
Acetophenone, 44
7-Acetoxybicyclo[2.2.1]heptadiene, 21
3-Acetoxytetracyclo[3.2.0.0^{2,7}.0^{4,6}]heptane, **21**
Acetylene dicarboxylic acid, mono-potassium salt, 95
5-Acetylnorbornene, 76
Adduct of 1,3,5-cyclooctatriene and dimethyl acetylenedicarboxylate, 41
Allyl cyclopropane, 31
Allylic hydroperoxides, 61
N-o-Aminobenzoyldiphenylamine, 91
2-Aminobiphenyl, 90
2-Aminopyridine, 46
    dimer of, **46**
Apparatus, photolytic, 11, 14
7-Azabicyclo[4.2.2]deca-2,4,9-trien-8-one, 67
Azabullvalene system, 69

Benzaldehyde, 56
2-Benzamidobiphenyl, 90
Benzenesulfonyl chloride, 37
Benzobicyclo[3.2.0]hept-2,6-diene, 24
Benzobicyclo[4.1.0]hept-2,4-diene, 24
Benzo[c]chrysenes, 57
Benzonorbornadiene, 23
Benzonorcaradiene, **23**
exo-2,3-Benzonorcaradiene-7-acetic acid, 92
exo-2,3-Benzonorcaradiene-7-carboxylic acid, 92
    chloride, 93
Benzophenanthrenes, 57
Benzophenone, 27, 49, 51, 73, 78
Benzopinacol, 52, 79
Benzo[f]quinoline-6-carbonitrile, **25**
2,3-Benzotricyclo[6.1.0.0^{4,9}]nona-2,6-dien-5-one, 27
1,2-Benzotropilidene, 23
3,4-Benzotropilidene, 24
N-Benzoylcarbazole, 91

Benzyne, 23, 27
Bicyclo[2.2.1]hepta-2,5-diene, 97
    2,3-dicarboxylic acid, 94
Bicyclo[3.2.0]hept-6-en-3-ols, 30
Bicyclo[3.2.0]hept-6-en-3-one, 28
Bicyclo[2.1.1]hexane, **31**
Bicyclo[2.1.1]hexan-2-one, 33, 88
Bicyclo[2.1.1]hex-2-ene, 35
Bicyclo[1.1.1]pentane, 35
Bicyclo[1.1.1]pentanes, 88
$anti$-1,3-Bishomocubanol, 85
1,1-Bishomocubyl mesylate, 86
3-(Bromomethyl)cyclobutyl bromide, 88
Bullvalene, 69
1,3-Butadiene, 39
N-n-Butylpyrrolidine, **36**

2-Chlorocycloheptanone, 54
Chlorosulfonyl isocyanate, 67
4-Cholestene-3-one, 42
Chrysenes, 57
$trans$-Cinnamic acid, 80, 103
Congressane, 82
Cuprous, bromide, 81
    chloride, 101
Cyanohydrin of cyclohexanecarboxylic acid, 54
Cyclobutene, 39
Cyclobutyldimethylamine oxide, 41
Cyclobutyl phenyl, carbinol, 88
    ketone, 88
Cyclobutyltrimethylammonium hydroxide, 41
1β,5-Cyclo-5β,10α-cholestan-2-one, **42**
3,5-Cycloheptadiene, 29
3,5-Cycloheptadienol, 29
3,5-Cycloheptadienone, 28
1,3,5-Cycloheptatriene, 30
Cyclohexanecarboxylic acid, 54
Cyclohexanone diethyl ketal, 72
cis,cis- and $cis,trans$-1,3-Cyclooctadiene, 44
1,5-Cyclooctadiene, 101

INDEX

Cyclooctatetraene, 67
1,3,5-Cyclooctatriene, 41
Cyclopentadiene, 23, 95

Diadamantane, 82
Diazomethane, 24, 93
Diazonorcamphor, 32
1,1-Dibromo-2(3-butenyl)cyclopropane, 77
Di-n-butylamine, 36
Di-n-butyl-N-chloramine, 38
Dicyclopentadiene, 95
endo-Dicyclopentadiene, 86
2,3-Dihydro-5,6-diphenylpyrazine, 50
Dimethyl acetylenedicarboxylate, 41
4,4'-bis(Dimethylamino)benzophenone, 49
2,3-Dimethyl-butan-2-ol, 60
2,3-Dimethyl-2-butene, 60
2,5 Dimethylfuran, 70
2,5-Dimethylfuran, endo-peroxide, 71
5,5-Dimethyl-2-methylenebicyclo[2.1.1]-hexane, 49
4,6-Dimethyl-α-pyrone, 99
5,5-Dimethyl-1-vinylbicyclo[2.1.1]hexane, 48
Di-α-naphthylethylenes, 57
trans-1,3-Dinitro-trans-2,4-diphenylcyclobutane, 79
trans-2,4-Diphenylcyclobutane-trans-1,3-dicarboxylic acid, 103
4,5-Diphenylimidazole, 50
4,5-Diphenyl-1-methylimidazole, **50**
4,4-Diphenyl-3-oxatricyclo[4.2.1.0^{2,5}]-nonane, **51**

Einstein, 2
Eosin Y, 61, 71
Ethyl, cyclohexanecarboxylate, 53
  diazoacetate, 93
  1-hydroxycyclohexanecarboxylate, **53**

Filters, 4, 7, 11, 14
p-Fluorobenzylmagnesium chloride, 56
3-Fluorophenanthrene, **55**
p-Fluorophenylmagnesium bromide, 56
cis- and trans-p-Fluorostilbene, 56
Formic acid, 58

2-Halo-3-hydroperoxy-2,3-dimethylbutane, 61
1,6-Heptadien-2-one, 35
1,2,6-Heptatriene, 77
1,5-Hexadiene, 31
cis,trans- and trans,trans-2,4-Hexadienoic acid, 58
3,4-Hexadienoic acid, **58**
1,5-Hexadien-3-ol, 33
1,4-Hexadien-3-one, 36
1,5-Hexadien-3-one, 33
cis, syn, cis- and cis, anti, cis-1,5,5,7,11,11-Hexamethyltricyclo[6.4.0.0^{2,7}]-dodecane-3,9-dione, 64
cis, syn, cis-5,5,7,8,10,10-Hexamethyltricyclo[6.4.0.0^{2,7}]dodecane-3,12-dione, 64
1-Hydroperoxide of methyl cyclohexanecarboxylate, 55
3-Hydroperoxy-2,3-dimethyl-1-butene, **60**
2-Hydroxyadamantane, 55
1-Hydroxycyclohexanecarboxaldehyde, 54
1-Hydroxycyclohexanecarboxylic acid, 53
α-1-Hydroxydicyclopentadiene, 85
syn-8-Hydroxydicyclopentadiene, 86

Isophorone, 63
  dimers, **62**
  head-to-head and head-to-tail dimers of, 62, 63

Lamp, choice of, 4
  mercury, 7, 9, 12, 13
  185 nm, 7
  resonance, 7
  visible source, 10
Light, output (watts), 8, 9
  wavelength distribution, 4
Lumicholestenone, **42**
6-epi-Lumisantonin, **65**

Maleic anhydride, 39
7-Methoxy-8-azabicyclo[4.2.2]deca-2,4,7,9-tetraene, 68
Methoxyazabullvalene, 67
3-Methoxy-4-azatricyclo[3.3.2.0^{2,8}]deca-3,6,9-triene, **67**
2-Methoxy-5-hydroperoxy-2,5-dimethyldihydrofuran, 70
Methylamine, 74
N-Methylaniline, 72
1-(N-Methylanilino)cyclohexane, 72
Methyl cyclohexanecarboxylate, 54

2-Methylenebicyclo[2.1.1]hexane, 35, 49, 77
Methylene Blue, 61, 71
3-Methylene-1,5-hexadiene, 77
cis-9-Methyl-1,2,3,4,4a,9a-hexahydrocarbazole, 73
trans-9-Methyl-1,2,3,4,4a,9a-hexahydrocarbazole, 72
Methyl 1-hydroxycyclohexanecarboxylate, 54
Methyllithium, 77
α- and β-Methylnaphthalenes, 24
N-Methyl-4-nitro-o-anisidine, 74
N-Methyl-5-nitro-o-anisidine, 75
3-Methyl-2-oxatetracyclo[4.2.1.0^{3,9}.0^{4,8}]-nonane, 76
9-Methyl-1,2,3,4-tetrahydrocarbazole, 73
Michler's ketone, 48
Myrcene, 48

Naphthalene, 23, 93
1-α-Naphthyl-2-β-naphthylethylenes, 57
β-Nitrostyrene, 79
trans-β-Nitrostyrene photodimer, 79
4-Nitroveratrole, 74
Norbornadiene, 82, 97
Norbornadienes to tetracyclo-[3.2.0.0^{2,7}-.0^{4,6}]heptanes, 22
Norbornene, 51
  dimer, 81
Norcamphor, 32

Outgassing, 16
Output from, commercial lamps, 9
  low pressure lamps, 8
Oxetane from photocycloaddition of benzophenone to norbornene, 51

Pentacyclo[4.4.0.0^{2,5}.0^{3,8}.0^{4,7}]decane-9,10-dicarboxylic acid anhydride, 83
anti-Pentacyclo[5.3.0.0^{2,5}.0^{3,9}.0^{4,8}]-decan-6-ol, 85
Pentacyclo[5.3.0.0^{2,5}.0^{3,9}.0^{4,8}]decan-6-one, 86
Pentacyclo[4.4.0.0^{2,5}.0^{3,8}.0^{4,7}]dec-9-yl methanesulfonate, 86
Pentacyclo[8.2.1.1^{4,7}.0^{2,9}.0^{3,8}]tetradeca-5,11-diene, 82
exo,trans,endo- and exo,trans,exo-Pentacyclo[8.2.1.1^{4,7}.0^{2,9}.0^{3,8}] tetradecane, 82
1,4-Pentadienes, 88
Peracetic acid, 53
3-Phenanthrylamine, 57
6-Phenoxyphenanthridine, 91
Phenylacetaldehyde, 56
2-Phenylbicyclo[1.1.1]pentanol-2, 87
2-Phenyldibenzo[d,f]-1,3-oxazepine, 91
1-Phenyl-4-penten-1-one, 88
6-Phenylphenanthridine, 90
  N-oxide, 89
5-Phenylphenanthridine-6[5H]-one, 89
2-Phenyl-3-(2-pyridyl) acrylonitrile, 25
Photochemical Wolff rearrangement, 92
Photon flux, 2, 3, 4, 5, 12
Photostationary state, 7
Picenes, 57
Pinacol, 60
Piperylene, 76
Power in watts, 2, 3

Quadricyclene, 97
Quantum yield, 2, 4, 5

Rayonet Preparative Photochemical Reactor, 12
Reaction details, 15
Relative intensity, 8
Rose Bengal, 60, 70

Safety, 16
α-Santonin, 65
6-epi-α-Santonin, 65
Sensitization, 10, 18
Silver nitrate complex of cis,trans-1,3-cyclooctadiene, 45
Sorbic acid, 58
Sunlight, 10, 96

Temperature control, 17
Tetracyclo[3.2.0.0^{2,7}.0^{4,6}]heptane, 97
Tetracyclo[3.2.0.0^{2,7}.0^{4,6}]heptane-2,3-dicarboxylic acid, 94
1,3,5,7-Tetramethylcyclooctatetraene, 99
Time, minimum conversion, 5
  of photolysis, 2, 4
endo, anti-Tricyclo[5.2.1.0^{2,6}]deca-4,8-dien-3-ol, 85
endo, syn-Tricyclo[5.2.1.0^{2,6}]deca-3,8-dien-10-ol, 86

Tricyclo[4.2.2.0^{2,5}]deca-3,7-diene-9,10-
  dicarboxylic anhydride, 83
Tricyclo[4.2.2.0^{2,5}]deca-7-ene-9,10-
  dicarboxylic anhydride, 84
Tricyclo[4.1.0.0^{1,3}]heptane, 77
Tricyclo[3.3.0.0^{2,6}]octane, **101**
Trimethyloxonium fluoroborate, 68
Triphenylene, 49

Triplet energies, 18
Tropone, 27, 29
α-Truxillic acid, **103**

1-Vinylbicyclo[2.1.1]hexane, 49

Wavelength spread, 7